IEE ELECTRICAL AND ELECTRONICS MATERIALS
AND DEVICES SERIES 2

Series Editor: N. Parkman

ELECTRICAL INSULATION

ELECTRICAL INSULATION

Edited by A. Bradwell

PETER PEREGRINUS LTD
on behalf of the
Institution of Electrical Engineers

Published by: Peter Peregrinus Ltd., London, UK.

© 1983: Peter Peregrinus Ltd

ISBN 0 86341 007 3

Printed in England by Short Run Press Ltd., Exeter

Contents

LIST OF CONTRIBUTORS

Chapter 1.
Dr. D.T.A. Blair
University of Strathclyde
George Street
Glasgow G1

Chapter 2
Dr. D. Binns
University of Salford
Electrical Engineering Dept
Peel Park
Manchester

Chapter 3
Professor R. Cooper
University of Manchester
Electrical Engineering Dept
Dover Street
Manchester

Chapter 4
Dr. N. Parkman
ERA Technology Ltd
Cleeve Road
Leatherhead
Surrey

Chapter 5
Mr. W. Reddish
167 High Street
Codicote
Stevenage

Chapter 6
Dr. H.M. Ryan
NEI Reyrolle Ltd
South Drive
Hebburn
Tyne & Wear NE31 1XA

Chapter 7
Mr. F.B. Waddington
Micanite and Insulators Ltd
GEC, Trafford Park
Manchester M17 1PR

Chapter 8
Mr. J. Heighes
Permali Gloucester Ltd
125 Bristol Road
Gloucester GL1 5SU

Chapter 9
Mr. J. Staight
ERA Technology Ltd
Cleeve Road
Leatherhead
Surrey

Chapter 10
Mr. R.C. Blatcher
GEC Switchgear
Lichfield Road
Stafford

Chapter 11
Mr. A. Kallinikos
GEC Power Engineering
Micanite and Insulators Ltd
Trafford Park
Manchester 17

Chapter 12
Mr. G.C. Stevens
Central Electricity Research
Laboratories
Kelvin Avenue
Leatherhead
Surrey KT22 7SE

Chapter 13
Mr. C. Clemson
GEC Distribution Switchgear
Higher Openshaw
Manchester M11 1FL

Chapter 14
Mr. W.P. Baker
Electricity Council Research
Laboratory
Capenhurst
Cheshire

Chapter 15
Mr. A.W. Stannett
Coordinator
Central Electricity Research
Laboratories
Kelvin Avenue
Leatherhead
Surrey KT22 7SE

Foreword

This Vacation School in Electrical Insulation initiated from the regular meetings of the IEE Professional Group S2 - Materials during the chairmanship of Professor Cooper. It was felt that there was a declining number of courses for students in power engineering, particularly with respect to electrical insulation, and that a post-graduate course would be welcomed. A small working group was set up consisting of Professor R. Cooper, D. O'Sullivan, F.W. Waddington, J.W. Wood and myself to consider the content of the course. To be of value in training Insulation Engineers, it was agreed that the course should include aspects of the fundamental properties of dielectrics, their desirable properties for particular power applications and the means of assessing the performance of materials both in the laboratory and in service. It would also be necessary to cover the range of insulants, gases, liquids and solids, and to consider these areas where gas/solid and liquid/solid interfaces react. Fifteen U.K. authors were then invited to condense their expertise in these diverse areas of electrical insulation into brief lectures with accompanying notes.

Subsequent S2 committees, under the chairmanships of F.W. Waddington and N. Parkman, endorsed and encouraged the course and suggested that the course notes be presented in book form, since no other book was available to students covering these aspects of electrical insulation.

As editor it has been my task not to referee the authors contributions but to outline the general subjects to be covered to prevent duplication and to try to bring continuity and completeness to what is essentially a series of lectures. The space allowed each author was small (approximately 7,000 words) and this has restricted detailed discussions. However there are copious references to the original published work. I have enjoyed reading the texts during the editing and hope that students will similarly find stimulus both in the book and in the lecture course to delve further into this interesting field of Electrical Insulation.

I offer my thanks to the other members of the organising committee, but particularly to the authors for delivering such succinct yet comprehensive texts within the timescales allowed.

British Railways Technical Centre   Alan Bradwell
London Road
Derby.   June 1983

Chapter 1

Breakdown in gases

D. T. A. Blair

1.1 FUNDAMENTAL PROCESSES AT THE MOLECULAR LEVEL

When an electric field is applied in a gas-filled region, no current will flow (apart from capacitance current) unless free electric charges are present. When the applied field strength is low, the current flow is due entirely to charged particles produced by external means (background radioactivity and cosmic radiation), and is usually a very small fraction of a microampere. At higher values of electric field strength, inelastic electron-molecule collisions occur, causing changes in the states of the orbiting electrons in the molecules: some of these inelastic collisions provide further charged particles and can eventually make the gas highly conducting. We shall consider the most important of the inelastic collision processes in turn.

Excitation. In excitation, an orbiting electron in the molecule is raised to a higher energy level than it occupies when the molecule is in its ground state. The excited state is not stable and excitation is always followed by another process that uses the excess energy possessed by the excited molecule. Often the excited molecule loses its excess energy by emitting a photon. These processes may be represented:

excitation $\qquad M + e^- \rightarrow M^* + e^-$;

photon emission $\qquad M^* \rightarrow M + h\nu$.

Here e^- represents the electron, M the molecule with which it collides, M^* the excited molecule, and $h\nu$ a photon of frequency ν and energy $h\nu$, h being Planck's constant. When the gas is not monatomic, dissociation may occur along with excitation. The lifetime of an excited state is usually of the order 10^{-8}s or less, but in some gases there occur so-called metastable states, which have lifetimes up to about 10^{-2}s. De-excitation processes other than photon emission also occur.

Collisional Ionization. If an electron possesses kinetic energy at least equal to the ionization energy of the molecule with which it collides, then the molecule can be ionized in the collision:

$$M + e^- \rightarrow M^+ + 2e^-$$

The positive ion may be produced originally in an excited state, and enter the ground state after emitting a photon. If the gas is not monatomic, dissociation may occur along with collisional ionization, and one or more of the dissociation products may be produced in an excited state. In any case, an additional free electron is released into the gas.

<u>Photoionization</u>. When a photon has an energy $h\nu$ at least equal to the ionization energy of a gas molecule, it can ionize the molecule:

$$M + h\nu \rightarrow M^+ + e^-.$$

<u>Attachment.</u> If a gas molecule has unoccupied energy levels in its outermost group, then a colliding electron may take up one of these levels, converting the molecule into a negative ion:

$$M + e^- \rightarrow M^-$$

In this process, some means must exist to carry off the excess energy. If the molecule is not monatomic then dissociation may occur along with attachment, and in any case the negative ion may initially be produced in an excited state. Alternatively, a three-body collision may occur, in which the excess energy is converted to kinetic energy in the third colliding particle.

Ionization and attachment are competing processes in the sense that ionization increases the number of free electrons present in the gas, while attachment decreases it.

<u>Detachment.</u> This occurs when a negative ion gives up its extra electron, and becomes a neutral molecule. The electron is again free, and therefore able to ionize by collision. At its simplest, the process may be represented:

$$M^- \rightarrow M + e^-$$

Energy is required to separate the electron from the negative ion, and this may be provided by another particle (collisional detachment), by a photon (photodetachment) or by other means.

1.2   THE ELECTRON AVALANCHE

Suppose a free electron is released by some external effect (background radioactivity or cosmic radiation) in a gas where an electric field exists. If the field strength is high enough, and if the electron is not captured in an attaching process, then the electron is likely to ionize a gas molecule by collision. There will then be two free electrons and a positive ion. The two electrons will both be able to cause further ionizations by collision, provided they do not undergo attachment, and there will then be four electrons and three positive ions. The process is cumulative, and the number of free electrons will go on increasing as they continue to move under the action of the electric field. The swarm of electrons and positive ions produced in this way is called an electron avalanche. In the space of a few millimetres it may grow until it contains many millions of electrons.

The formation of an electron avalanche is the first event (after the release of an initiating electron) in gas insulation failure. The avalanche continues to grow until the electrons either reach the positive electrode or travel into a region where the electric field strength is not high enough to support significant collisional ionization.

For any given avalanche path, the numbers of electrons and ions in the avalanche will vary statistically from one avalanche to another. The reason for the statistical variation lies in the random nature of the collision processes involved.

1.3 THE TOWNSEND THEORY OF BREAKDOWN

In a breakdown by the Townsend mechanism, the first or Primary electron avalanche usually has to be initiated by an external factor, such as background radioactivity. The processes occurring in this avalanche initiate one or more new avalanches, called secondary avalanches, and these in turn initiate further secondary avalanches. An unbroken chain of avalanches constitutes a self-sustaining discharge.

The electrons that initiate secondary avalanches may be produced by several different processes. For example, they may be produced at the cathode by the impact of positive ions from a previous avalanche, by the impact of photons from exciting collisions that took place in a previous avalanche, or by the impact of metastable molecules that have a sufficiently long lifetime to diffuse to the cathode before they revert to the ground state.

Because of the statistical nature of the processes involved, not every externally produced electron will succeed in triggering a self-sustaining discharge. However, if the applied electric field remains sufficiently high, it is only a matter of waiting long enough until one does.

As a result of the rapid succession of avalanches, the number of charged particles in the gap increases with time, and the subsequent course of events is governed by the space-charge field associated with these charged particles. In an initially uniform electric field, the build-up of charge leads to an unstable situation where the current increases very rapidly while the applied voltage collapses. In a highly nonuniform field, the charges can form a kind of electrostatic shield around the electrode where the electric field strength is greatest, and therefore cause quenching of

the self-sustaining discharge: the resulting corona discharge may be either pulsating or continuous, and the
average current may be a considerable fraction of an ampere.

1.4 THE CRITERION FOR TOWNSEND BREAKDOWN

Consider a discharge gap consisting of two metal
electrodes immersed in a gas, with direct voltage applied
between them, giving an electric field that does not vary
with time. Suppose that in each second n_0 free electrons
are released in the gas, and to keep the argument simple,
let them be produced at the surface of the cathode (that is
the negative electrode). Under the action of the electric
field, these electrons will travel across the gap to the
anode, and if the field strength is high enough they will
produce electron avalanches. By choosing n_0 sufficiently
large, or by considering the process to continue over a
sufficiently long period, we can make the effects of statistical variations between individual avalanches negligible,
and we are justified therefore in arguing as though all the
avalanches were identical.

Now, if Mn_0 electronic charges arrive at the anode when
only n_0 left the cathode, then $(M - 1) n_0$ must have been
released in the gas by collisional ionization. Consequently,
$(M - 1)n_0$ ionizing collisions must have occurred, and
$(M - 1)n_0$ positive ions will eventually reach the cathode.
Here we make the reasonable assumptions that an ionizing
collision can release only one additional free electron,
leaving behind a singly charged positive ion, and that we
may neglect recombination between electrons and positive
ions.

We define a secondary coefficient γ as the (average)
number of secondary electrons released at the cathode per
ionizing collision in a single avalanche. As a direct
result of n_0 avalanches, therefore, the number of secondary
electrons released at the cathode is $\gamma (M - 1)n_0$.

These $\gamma(M - 1)n_0$ electrons are only the first generation of secondary electrons: they will in turn give rise to
more avalanches, producing a further $[\gamma(M - 1)]^2 n_0$ secondary electrons. The process will continue, and the total
number of electrons emitted from the cathode each second,
including all secondaries, will be

$$n_0(1 + \mu + \mu^2 + \mu^3 + \ldots)$$

where $\mu = \gamma(M - 1)$. If μ is equal to or greater than unity,
the series will diverge, and the current flowing in the gap
will correspondingly increase without limit. The criterion
for onset of a self-sustaining discharge is therefore

$$\gamma(M -1) = 1 \qquad\qquad (1.1)$$

This criterion is valid for a wide range of conditions.
We have made few assumptions about the physical processes

that govern the numerical value of M: obviously collisional ionization must be involved, but equation (1.1) remains the same if this is accompanied by attachment, detachment and/or other processes. Further, the criterion is clearly valid for any secondary process that causes secondary electrons to be produced at the cathode surface, and not only for the three specific secondary processes that we mentioned. For a process that causes secondary electrons to be produced elsewhere (for example photoionization of the gas), this criterion is not strictly valid, but it provides a good approximation in many cases.

To obtain equation (1.1) in a form that is useful in a specific case, we need to substitute an appropriate expression for M. We must therefore consider how avalanches grow in particular cases.

1.4.1 Electron avalanche growth in nonattaching gases.

We define a coefficient α for ionization by collision such that, when an electron avalanche progresses an elemental distance dx along its path, the number of ionizing collisions per electron is $\alpha\, dx$. If the avalanche contains $n_e(x)$ electrons at x, then this number will increase by $dn_e(x)$ as the avalanche progresses from x to $(x + dx)$, where

$$dn_e(x) = n_e(x)\, \alpha\, dx \qquad (1.2)$$

If the avalanche begins at x = 0, then the number of electrons it contains when it reaches x = s is obtained by rearranging equation (1.2) and integrating:

$$n_e(s) = n_e(0)\, \exp \int_0^s \alpha\, dx \qquad (1.3)$$

If we take s = d, the length of the gap between the cathode and the anode measured along the avalanche path, then we may identify $n_e(0)$ with n_0 and $n_e(d)$ with Mn_0. We may therefore write

$$M = \exp \int_0^d \alpha\, dx \qquad (1.4)$$

For the particular case where the electric field is uniform, this reduces to

$$M = \exp(\alpha d) \qquad (1.5)$$

1.4.2 Electron avalanche growth in attaching gases.

We shall consider the case where only stable negative ions are formed, that is no detachment occurs. (The case of a gas that forms both stable and unstable negative ions has been analysed, but will not be considered here).

By analogy with the ionization coefficient α we define an attachment coefficient η. Each ionizing collision causes the number of free electrons to increase by one,

while each attaching collision causes it to decrease by one.
We obtain instead of equation (1.2)

$$dn_e(x) = n_e(x) \; \bar{\alpha} \; dx \tag{1.6}$$

where $\bar{\alpha} = \alpha - \eta$ (sometimes called the effective ionization
coefficient). In place of equation (1.3) we obtain, with
s = d,

$$n_e(d) = n_e(0) \; \exp \int_0^d \; \bar{\alpha} \; dx \tag{1.7}$$

for the general case, and

$$n_e(d) = n_e(0) \; \exp (\bar{\alpha}d) \tag{1.8}$$

for the uniform-field case.

To evaluate M we must consider the total number of
electronic charges (free electrons plus negative ions)
reaching the anode. This leads to

$$M = (\alpha/\bar{\alpha}) \; \exp(\bar{\alpha}d) - (\eta/\bar{\alpha}) \tag{1.9}$$

for the uniform-field case.

1.5   PASCHEN'S LAW

1.5.1   Uniform electric fields.

For a given gas, the value of the ionization coefficient
α must obviously depend on the electric field strength. It
is fairly easy to show that, if collisional ionization is
always a two-body process, then α/N is a function of E/N
only, where E is the electric field strength and N is the gas
number density.

For a nonattaching gas in a uniform electric field, we
may combine equations (1.1) and (1.5) and write the breakdown
criterion as

$$\gamma[\exp (\alpha d) - 1] = 1 \tag{1.10}$$

If we write

$$\alpha/N = f(E/N) \tag{1.11}$$

and make the assumption that we can also write γ as a function of E/N only, say

$$\gamma = g(E/N) \tag{1.12}$$

then, writing the values of electric field strength and
applied voltage at breakdown as E_s and V_s respectively, we
obtain for the breakdown criterion

$$g(V_s/Nd) \; [\exp\{Nd \; f(V_s/Nd)\} - 1] = 1 \tag{1.13}$$

(For any value of Nd, there is only one value of V_S that will satisfy this equation. In other words, the value of the breakdown voltage V_S is determined uniquely by the product of gas density and gap length. This is Paschen's law.) It applies also to attaching gases, as may be shown quite easily.

The breakdown voltages of uniform-field gaps in a given gas can therefore be shown conveniently on a plot of V_S against Nd.   This plot is called a Paschen curve, and has the general form sketched in Fig. 1.1a.   Since each value of Nd gives a unique value of V_S, it must also give a unique value of V_S/Nd.   This latter quantity is equal to E_S/N, and the Paschen curve can equally well be drawn as a plot of E_S/N against Nd.   This takes the general forms sketched in Fig. 1.1b.   Here the case of attaching gases is distinguished by the occurrence of a limiting value of E/N, below which uniform-field breakdown is not possible for any finite gap length.   (Strictly, the existence of a limiting E/N depends on the formation of stable negative ions, either directly by collisional attachment or by the formation of unstable negative ions that are subsequently stabilised in another collision process).

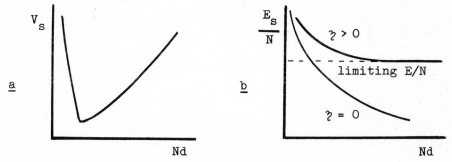

Fig. 1.1   Forms of Paschen curves

The limiting E/N exists in an attaching gas because there is some value of E/N below which attachment predominates over collisional ionization, and avalanche formation is therefore not possible.   In a gas that forms only stable negative ions, the limiting E/N is for practical purposes the value of E/N at which $\alpha = \eta$.

Fig. 1.1a shows the existence of a minimum breakdown voltage.   This is usually of the order of 100 V, and occurs typically at an Nd value corresponding to a gap length of 10 mm at a gas pressure of about 5 torr at normal ambient temperature.

To the right-hand side of the minimum, breakdown voltage increases with increasing gas pressure.   This is basically because an increase in gas pressure leads to a decrease in the electron mean free path: a higher electric field strength is therefore required to allow some of the electrons to pick up the ionization energy from the field between collisions. For a uniform-field gap of length 10 mm, the breakdown field strength at atmospheric pressure and normal ambient temperature is about 3 kV/mm for air and about 9 kV/mm for SF_6.

To the left-hand side of the minimum, the breakdown voltage increases with decreasing gas pressure. In this region, electron-molecule collisions are relatively infrequent, and a decrease in gas pressure makes avalanche formation more difficult by creating a shortage of gas molecules for the electrons to collide with.

1.5.2 Nonuniform electric fields.

In a nonuniform electric field, the discharge becomes self-sustaining at a voltage V_{si}, and the self-sustaining discharge can be either a complete breakdown or a corona discharge. Here Paschen's law has to be replaced with the more general similarity principle, according to which V_{si} is a function of Nd only, provided that changes in d are accompanied by proportional changes in all other gap dimensions.

In arrangements where the field nonuniformity is very marked, the onset of a self-sustaining discharge is governed by events in a small region where the electric field strength is at its highest. In such cases, Paschen's law may be replaced, to a useful approximation, with the rule that $E_{si}r$ is a function of Nr only, where E_{si} is the highest value of electric field strength in the gap when the applied voltage is V_{si} and r is the radius of curvature at the electrode surface where the highest field strength occurs. This rule is justified by the similarity principle, but strictly only where changes in r are accompanied by proportional change in all other gap dimensions.

1.5.3 Deviations from Paschen's law.

Marked deviations from Paschen's law (or the similarity principle) are often observed at gas pressures of a few atmospheres, involving reduction of the breakdown voltage to as low as 50 per cent of the Paschen values. In many cases, and certainly in SF_6 under conditions of practical interest, these reductions are most probably caused by distortion of the electric field due to roughness of the electrode surfaces[1.1] or by the presence of solid particles, either fixed or free to move[1.2].

1.6 THE STREAMER THEORY OF BREAKDOWN

The streamer theory was introduced initially to explain breakdown in the apparent absence of secondary processes. There are two requirements for the initiation and propagation of a streamer:

1.   Photons from the avalanche head produce free
      electrons in the gas by photoionization (or
      perhaps by photodetachment from negative ions);

2.   The space charge produced in the avalanche causes
      sufficient distortion of the electric field that
      those free electrons move towards the avalanche
      head, and in so doing generate further avalanches
      in a process that rapidly becomes cumulative.

The electrons in the avalanche can be assumed to have a
roughly spherical distribution, the radius of the sphere
being determined by diffusion. As the electrons advance
rapidly, the positive ions (and possibly negative ions) are
left behind in a relatively slow-moving tail.

The field will be enhanced in front of the head of the
avalanche, and some of the field lines from the anode will
terminate in this head (Fig. 1.2). Behind the head of the
avalanche, the field between the electrons and the ions is
in the opposite direction to the applied field, and the
resultant field strength here is less than the applied field
strength. Still further back, the field between the cathode
and the positive ions is enhanced again.

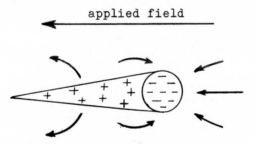

Fig. 1.2   Avalanche space-charge field

The space-charge distortion of the field begins to have
a detectable effect on the avalanche for electron numbers of
the order of 10^6. For numbers of the order of 10^8, the
space-charge field is comparable with the applied field, and
streamer propagation is possible, according to this theory.

It was claimed in the past that streamer theory was
necessary to explain breakdown in uniform electric fields at
high Nd values, but more recent work indicates that such
breakdown can usually be explained satisfactorily in terms
of Townsend theory. In a case where the cathode is a long
way from the head of a developing avalanche, as for example
in a lightning discharge, it seems unlikely that a cathode-
dependent secondary process is active, and here the discharge development is more easily visualised using the
streamer-theory concepts. In any case, if we accept photoionization as a possible secondary mechanism in Townsend
theory, and allow that space charge may enter into the

breakdown criterion, it becomes difficult to see any
fundamental difference between the two theories.

1.7   PREDICTION OF BREAKDOWN VOLTAGES

One of the obvious practical applications of gas breakdown theories is to predict the breakdown voltage of any
given arrangement using gas insulation.

Any attempt to predict breakdown voltages on the basis
of Townsend theory depends on a knowledge of the appropriate
secondary coefficient, and generally this knowledge is not
available.   In fact, most techniques for breakdown-voltage
prediction are based on streamer theory, usually with a
rather simple assumption regarding the criterion for
streamer propagation. Usually this is equivalent to using
Townsend theory and assuming an appropriate value for the
secondary coefficient.

The simplest breakdown criterion is obtained by assuming
that streamer-induced breakdown occurs when the number of
electrons in the average avalanche reaches some critical
value, often taken as 10^8.   For a nonattaching gas, the
number of electrons in the average avalanche is given by
equation (1.4); for an attaching gas that forms only stable
negative ions, it is given by equation (1.7) with $n_e(0) = 1$.
In either case the intergration has to be performed along
the field line that will give the greatest electron number.
In the case of the attaching gas, if there is a position on
this line where $\bar{\alpha} = 0$, then the integration has to be carried out between this position and the position where the
electric field strength (and hence $\bar{\alpha}$) is greatest. Application of this kind of criterion gives results in accord with
the similarity principle, provided that α/N and (where it
exists) η/N are functions of E/N only.

Pedersen has discussed in detail the question of setting up a useful streamer-breakdown criterion with
particular reference to air[1.3] and SF_6[1.4].

Where corona occurs, these techniques predict the
corona inception voltage, not the breakdown voltage. For
insulation design purposes, it is the corona inception
voltage that is usually required, because in most applications corona is to be avoided. (Prediction of the breakdown
voltage in these cases is not easy).

While these prediction methods give useful results
under many conditions, they fail in the region where large
deviations from Paschen's law occur (Section 1.5.3, above).

The breakdown characteristics of gases over a wide
range of conditions have been reviewed by Blair[1.5].

1.8   ELECTRODE EFFECTS

Townsend theory suggests that the nature of the cathode

material may affect the breakdown voltage of gas insulation by influencing the value of the secondary coefficient γ. However, at pressures up to at least atmospheric, changing the cathode material usually changes V_S by less than about 1 per cent, except in the region close to the Paschen minimum.

Simplified streamer criteria, based on the number of electrons in the average avalanche, take no account of electrode material.

At gas pressures of a few atmospheres, the breakdown voltage is sometimes found to depend on the nature of the cathode surface, and the effect is probably due to the presence of small protrusions on the surface. Here a conditioning effect often occurs, repeated breakdown of the gas causing the breakdown voltage to rise gradually until a plateau value is reached. The extent of the conditioning effect is generally unpredictable. (Paschen's law fails).

1.9 CORONA AND BREAKDOWN IN NONUNIFORM ELECTRIC FIELDS

In a uniform-field spark gap, when the conditions have been fulfilled for breakdown, a highly ionized conducting channel propagates rapidly across the gap until it bridges the electrodes, and breakdown ensues. In a highly nonuniform-field gap, the conditions for breakdown may be fulfilled within a region where the applied electric field strength is high, but the ionized channel may not be able to propagate through the region where the field is lower. In such cases a corona discharge occurs. This can be illustrated by reference to the voltage-current characteristics sketched in Fig. 1.3 (where the scales cover many orders of magnitude).

Fig. 1.3a shows the kind of characteristic that is usually obtained for a uniform-field gap. In region I the current is due entirely to external effects, such as background radioactivity, cosmic radiation or other irradiation of the gap. The current increases with increasing voltage because not all of the charge carriers produced are collected at the electrodes, some being lost in electron-ion recombination and/or back-diffusion of electrons to the cathode. In region II the electric field strength is high enough to ensure collection of all the charge carriers produced in the gap, but too low to cause ionization by collision. Region III is the Townsend region, where amplification of the gap current takes place as a result of collisional ionization and secondary processes. The upper limit of region III is set by breakdown at the voltage V_S.

Fig. 1.3b shows the kind of characteristic that is usually obtained for a nonuniform-field gap with a single region of high electric field strength in which corona occurs. Regions I, II and III have the same significance as in Fig. 1.3a, but now the upper limit of region III is set by corona inception at the voltage V_i. The effect of corona is to inhibit breakdown until the higher voltage V_S is reached.

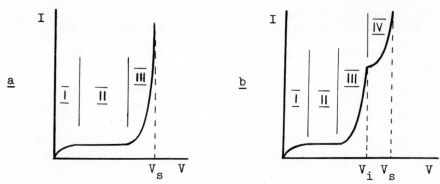

Fig. 1.3   *Current-voltage characteristics*

Fig. 1.4 illustrates the kind of behaviour that is typical for a nonuniform-field gap in an attaching gas. For a fixed gap length, corona precedes breakdown at gas pressures p below some critical value p_c, and for higher pressures the curve of sparkover voltage V_s against p is a continuation of the curve of corona inception voltage V_i against p.

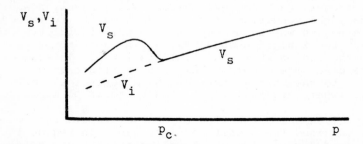

Fig. 1.4   *Characteristic of a highly nonuniform-field gap in an attaching gas*

1.10   BREAKDOWN TIME LAGS

Consider firstly an arrangement where corona does not occur. If a voltage in excess of V_s is applied suddenly, breakdown does not take place immediately, but only after a time lag. This time lag is the sum of two components: a statistical time lag and a formative time lag.

The statistical time lag is the time that elapses between the application of the voltage and the occurrence of the free electron that initiates the breakdown. Because of statistical effects in the fundamental breakdown processes, the free electron that initiates the breakdown is not necessarily the first free electron that occurs after voltage application. Consequently, the mean statistical time lag may be considerably greater than the mean time between the occurrence of free electrons.

The formative time lag is the interval between the occurrence of the initiating electron and the collapse of the applied voltage. Theoretical computations of the formative time lag are complicated by the need to consider in detail the effect of space charge on the development of the breakdown process.

In an arrangement where corona can occur before breakdown, the situation is more complex. If the formative time of the corona is long compared with the rise time of the applied voltage, then the corona cannot have the shielding effect that it has with dc or power-frequency voltages, and impulse breakdown can sometimes occur at a voltage below V_s.

Breakdown time lags are an important practical consideration when a protective spark gap is placed in parallel with a piece of apparatus (such as a power transformer) whose breakdown is to be avoided.

1.11 LOW PRESSURE AND VACUUM BREAKDOWN

Fig. 1.1a indicates that, if a uniform-field gap of fixed length is operated at low pressure, its breakdown voltage increases with decreasing pressure. The increase continues, although not always monotonically, into the vacuum region.

At a gas pressure of 10^{-4} torr, the length of the electron mean free path is of the order of a metre, so in a gap of the order of millimetres any electron that is present will probably cross the gap without making even a single collision with a gas molecule. Breakdown therefore cannot take place by an avalanche mechanism of the kind we have discussed above.

Several different mechanisms of vacuum breakdown have been proposed. All of these mechanisms involve the release of a gas or vapour from the electrodes, the eventual breakdown taking place in this gas or vapour.

Vacuum breakdown is characterised by high values of electric field strength, typically 30 kV/mm for a gap of 10 mm length.

Breakdown voltages in vacuum show strong conditioning effects, in that repeated measurements of breakdown voltage give gradually increasing values until a plateau value is reached. The cathode material has a large effect, with a hard material (such as stainless steel) giving a conditioned breakdown voltage as much as three times as high as a soft material (aluminium or copper). Surface contamination of the electrodes is also important.

A useful introductory review of vacuum breakdown is given by Van Oostrom and Augustus[1.6]. A more detailed review is given by Chatterton[1.7].

REFERENCES

1.1   Pedersen, A., 1970, 'The Effect of Surface Roughness on
      Breakdown in SF_6', IEEE Trans., PAS-94, 1749-1753.

1.2   Cookson, A.H., Bolin, P.C., Doepkin, H.C., Wootton, R.
      E., Cooke, C.M. and Trump, J.G., 1976, 'Recent Research
      in the United States on the Effect of Particle Contam-
      ination reducing the Breakdown Voltage in Compressed
      Gas-Insulated Systems', CIGRE, paper 15-08.

1.3   Pedersen, A., 1967, 'Calculation of Spark Breakdown and
      Corona Starting Voltages in Nonuniform Fields', IEEE
      Trans., PAS-86, 200-206.

1.4   Pedersen, A., 1970, 'Criteria for Spark Breakdown in
      Sulfur Hexafluoride', IEEE Trans., PAS-89, 2043-2048.

1.5   Blair, D.T.A., 1978, 'Breakdown Voltage Characteristics',
      in 'Electrical Breakdown of Gases' (Meek, J.M. and
      Craggs, J.D., editors), Wiley, 533-653.

1.6   van Oostrom, A. and Augustus, L., 1982, 'Electrical
      Breakdown between Stainless-Steel Electrodes in Vacuum',
      Vacuum, 32, 127-135.

1.7   Chatterton, P.A., 1978, 'Vacuum Breakdown', in
      'Electrical Breakdown of Gases (Meek, J.M. and Craggs,
      J.D., editors), Wiley, 129-208.

Chapter 2

Breakdown in liquids

D. F. Binns

2.1 INTRODUCTION

The fundamental electrical conduction and breakdown
processes that can occur in insulating liquids are of importance to the electrical power industry. There are a
large number of liquids in use and their purity is usually
ill defined. The electrodes used to investigate their
electrical characteristics and to make routine tests on
samples are of various materials and surface conditions
which will affect the behaviour. Processes of charge carrier multiplication that may control breakdown of carefully
prepared laboratory samples in which high voltage gradients
can be maintained will not occur in contaminated liquids
which rapidly break down at low stress.

The most used and important liquid insulant is mineral
transformer and switchgear oil that meets a specification
such as B.S.148(2.1)or IEC 296(2.2). Mineral oil is relatively cheap and has acceptable electrical strength and heat
transfer properties. It has a proven record over 80 years
or so for use with solid insulation at relatively low design
stresses up to 16kVrms/mm, albeit in very high voltage equipment. Mineral oil contains many component liquids, dissolved
gases and particles with the addition of oxidation inhibit-
ers and it is not of precisely controlled composition. Once
a transformer core and windings have been immersed in oil
that fills the tank and radiators the oil is contaminated
with particles, gas and water that can never be completely
removed by filtration, degassification and drying. Liquid
insulants are more difficult to maintain in good condition
than gases. The contamination of the gas that is used in a
large metal-clad substation can be controlled so that the
gas behaves almost like a very clean laboratory sample; any
comparable processing of bulk liquids is impracticable.

The flash point of mineral oil can be as low as 140°C
as specified in B.S. 148 (2.1)and it will burn readily if it
is at the heart of a fire. Furthermore, electrical discharges in the oil can produce explosive gases such as hydrogen that may accumulate in the tank or in a switchroom.
Hence there is an interest in the various low flammability,
non-combustible or explosion-free liquids including silicones and esters that are replacing polychlorinated biphenyls
for low fire risk. Non-combustible fluids that vaporize in

equipment at temperaturesranging mainly from 50°C to 150°C at 1 bar pressure, such as fluorocarbons or freons, are also being considered as insulants. Vapour bubbles will stream through these liquids near hot spots and consequently the electrical strength of the vapour is also of interest. Such liquids have also been pumped as a spray over transformer windings where they vaporize under hot conditions and partially displace an associated insulating gas such as SF_6 (2.3). Other liquids of interest include liquid nitrogen, liquid hydrogen and liquid helium which could serve as both insulant and coolant in cryogenic and superconducting cables should these ever be adopted. Mention should also be made of mineral oils and other liquids formulated for use in cables and capacitors. It should be noted that the various physico-chemical differences between liquids ensure that each has a unique insulation characteristic.

2.2. THE STREAMER BREAKDOWN PROCESS

The mechanisms of breakdown in relatively uncontaminated liquids subjected to impulse voltages are now thought to bear many similarities to the classical processes in gases (see Chapter 1). Electrons leaving the cathode and additional ones present in the liquid gain energy in the electric field as they move towards the anode. They collide with molecules of the liquid and if the voltage gradient is high enough dissociation of the liquid can take place to form micron-sized gas pockets or channels. These lower density regions, created essentially by energy dissipated in them, form most easily adjacent to the cathode surface but they may also form within the bulk liquid, especially in a uniform field parallel plane gap. Within low density regions Townsend-type avalanche processes may occur causing a multiplication of charge carriers; some charge carrier multiplication may also take place in the liquid. As electrons are about 10^5 times as mobile in a liquid as ions, positive ion accumulation can occur. If a series of very small low density regions are produced in a parallel plane gap a chain of them could be supplied with electrons from a favourable emission site on the cathode. As this discharge path becomes established along a slightly jagged track following the low density nuclear sites and running away from the cathode, the breakdown process suddenly changes to a new mode of development resembling a streamer process in a gas. This streamer process may arise when a sufficiently high charge density, probably of positive ions, is established at the head of the primary discharge rather as would occur in a gas. Several streamers develop occupying a wide conical volume and moving towards the anode. Local avalanches running in gas channels create a positive space charge that pushes the streamer still farther forward in a progressive movement.
With a negative point the breakdown process is similar except that only a very short discharge track needs to be created since the point is almost able to initiate streamer development itself. With positive point, electron

avalanches move towards the point initially forming gas
channels adjacent to the electrode surface in which charge
multiplication can occur. Positive space charge is built
up as electrons are drawn off into the point, the positive
charge acting like an extension of the point and causing
positive streamers to propagate towards the cathode. These
streamers progressing from anode to cathode are composed of
a series of avalanches moving in the opposite direction into
the advancing positive charge region and extending the
streamer in steps.

2.3 OPTICAL EXPERIMENTAL TECHNIQUES

The development of a low density region can readily be
observed using the techniques of schlieren photography or
the shadowgraph developed progressively by many workers (2.4)
(2.5),(2.6),(2.7). These valuable experimental techniques can
reveal minor details of the developing discharge pattern
and allow them to be related to the corresponding transient
current and light emission data. Referring to Fig 2.1

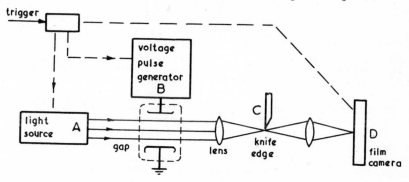

Fig.2.1 Elements of a schlieren system

which shows the elements of a schlieren system, the light
source at A can be a flash tube and collimating lens providing an intense parallel light beam or alternatively can be
a laser. The beam passes through the liquid between the
electrodes at B before being focussed on to a knife edge at
C. The knife edge can be moved with a micrometer so as to
cut off all or part of the light that would pass C. Any
light that might skirt round the knife edge would be focussed on to a still or moving film at D. If the knife edge is
critically adjusted so that it just cuts off all the light,
then when a discharge occurs in the gap and there is a
consequent temperature and density change in those parts of
the liquid affected by the discharge, the change in refractive index will cause the beam to be deflected. It may then
pass beneath the knife edge and be focussed at D. The
effect is as if the low density regions emit light, whereas
in fact the light is derived from the illuminating beam and
refracted by the discharge. The very feint light emitted

by the discharge itself can be separately though less easily
recorded on film using image intensification. Preferably
synchronisation is provided of the voltage pulse developed
at B, the light pulse at A and the camera at D. Very high
speed cameras, taking 10 to 20 pictures in 1 µs during discharge development, reveal the rate of advance of the discharge. This is easy to see for a negative point discharge
but more difficult for a positive point discharge which
propagates about 10 times as fast. The speed can also be
inferred from snapshots of discharges that have been arrest-
ed by a sudden removal of the applied voltage, the instant
at which this is done being varied in repeated runs of the
discharge. A limitation of optical methods is that they do
not reveal the very earliest stages of discharge development
and there is also some blurring of the image by movement of
the discharge envelope.

2.4 BREAKDOWN PHENOMENA

Various conclusions have been drawn about the breakdown
channel from the available evidence, especially the optical
data. Consider first the conduction processes that occur
near a cathode. Referring to Fig.2.2(a) and 2.2(b) the
cathode may be a point in a negative point/plane gap or a
plane in a negative plane/positive plane gap and in either
case similar manifestations occur. Forster (2.8) character-
ises both systems as diverging. As electrons leave the
vicinity of the point or a favourable location on the plane
cathode, under the action of a very high field, their
numbers may increase due to collisional processes as they
move towards the anode. In fact they appear to occupy a
narrow filamentary channel as they leave the plane cathode
until they have travelled a fraction of a mm where they
widen out into a discharge region as shown in Fig.2.2.(a).

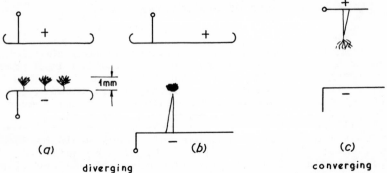

(a)   *(b)*   *(c)*

diverging   converging

Fig.2.2 Developing discharges in n-hexane

The region communicates with the cathode through a channel
like the trunk of a tree with the wide region looking like
the branched part of a tree. From this region negative
streamers propagate towards the anode at a typical speed of

0.2 to 1 mm/μs depending on the overvoltage applied. The length of the filament or trunk might be dictated by the need for the accumulation of charge to be far enough from the cathode to avoid disturbing the high voltage gradient at the filament source. Alternatively it might require such a path length to build up a sufficient charge density. On a plane cathode many tree-like processes develop alongside one another, as in Fig.2.2(a) whereas one or two filaments may leave a point cathode.

By contrast the positive point/plane arrangement is described as converging, Fig.2.2(c). The electrons available from the liquid or the plane cathode, or their descendants produced by some charge multiplication process, converge on the point anode. As the negative charge carriers approach the point they create suddenly a very high voltage gradient which produces prolific ionisation and leads to relatively violent positive streamers running back to the cathode.

The positive streamers appear to spread from anode to cathode. In fact they move in a series of discrete steps in the opposite direction. Each step is an electron avalanche, running into a positive space charge probably initiated by photoelectrons. A few positive streamers usually propagate together at a speed of several mm/μs, i.e. much faster than negative streamers. Positive point breakdown occurs at a lower voltage than for negative point in most liquids.

When positive and negative streamers have crossed the gap they leave continuous channels of lower fluid density which therefore have lower electrical strength and which also contain many charge carriers. A heavy current will then flow limited by the impedance of the external circuit, especially by the local capacitance associated with the immediate electrode structure.

The experimental observations of the formation of low density discharge regions, or bubbles, are consistent with the amount of energy given up to the liquid by charge carriers. As electrons move through the liquid under the action of a high field, of perhaps 50kV/mm intensity, their interactions with the liquid cause energy to be absorbed. This may cause dissociation of the liquid, typically requiring 4eV and producing hydrogen, and may also cause vaporisation (2.9).Using schlieren or shadowgraph techniques with laser illumination and image intensifiers to allow cameras to operate up to 2×10^7 frames/second discharges at or near the cathode are seen to develop into low density regions but the processes just prior to this are not observed. Neither are the processes near a plane anode although it has been claimed that these occur first and may prepare the way for cathode processes by emitting photons.

2.5 THE ROLE OF BUBBLES

The part played by bubbles in electrical breakdown of liquids varies according to circumstances. In degassed and dry liquids subjected to very high pulsed electric fields in the region of 50kV/mm and at ambient temperature or below copious electron production and multiplication at or near to the cathode can cause a bubble to develop directly as a result of the electrical energy input which raises the temperature to perhaps 50°C. In this case the bubble can be said to be a secondary manifestation of a developing discharge with the electron multiplication being the primary cause. On the other hand, liquids contaminated with gas and water and especially if heated cyclically will readily form bubbles from physico-chemical processes influenced by sustained alternating or direct voltages. In such cases the bubbles which are likely to form around sub-micron sized particles may be said to be the primary cause of discharge development at least in uniform fields.

When a spherical bubble of effective permittivity ε_R is subjected to a uniform electric field E imposed by electrodes far apart in comparison with the bubble diameter, then the field within the bubble is everywhere the same and equal to

$$\frac{3\varepsilon_R E}{2\varepsilon_R+1} \qquad \qquad \ldots \quad 2.1$$

The maximum value this can have is 1.5E, corresponding to $\varepsilon_R \to \infty$. For transformer oil with $\varepsilon_R = 2.2$ is 1.22E, for silicone fluid of $\varepsilon_R = 2.8$ it is 1.27E and for the ester Midel 7131 of $\varepsilon_R = 3.3$ it is 1.30E. This modest field enhancement in spherical bubbles must be considered alongside the greater ease with which ionisation can take place in a low density gas or vapour than in the liquid. Also as bubbles elongate in the field, especially under a.c.,the stress factor for the bubbles will decrease.

As discharges develop and progress from the vicinity of the cathode to the anode they take on a roughly spherical shape, or at least could be said to fit within a spherical envelope. If the space within such a sphere is highly conducting relative to the surrounding liquid then an effective sphere/plane geometry develops. With a point cathode/plane anode set at a gap 'g' the discharge region may approximate to a conducting sphere of diameter d which effectively reduces the gap to g-d, as in Fig.2.3. It may be shown that the field at the tip of a sphere facing the plane is a minimum approximately when d = 0.6g. Measurements of the speed of travel of the streamers away from a point cathode show a corresponding minimum when the discharge region has reached this diameter.

2.6 CONDITIONING

Measurements of breakdown voltage on samples of liquid tested between bare electrodes inevitably show variability. Repeated tests under nominally the same conditions can be made (a) with fresh liquid and fresh electrodes each time, (b) with fresh liquid each time but the same electrodes, (c) with fresh electrodes each time in the same liquid,

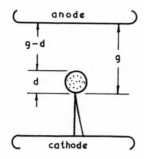

Fig.2.3   Representation of a spherical discharge
          region

(d) with the same electrodes and liquid each time. Method
(a) is necessary for testing silicone fluid when breakdown
debris in the liquid and attached to the electrodes can have
a very marked effect on breakdown voltage. The electrodes
can be cleaned after each breakdown and the silicone fluid
then filtered or discarded. Method (d) is the usual method
of test and if the bulk of liquid in the test vessel is very
large in relation to the gap volume that is contaminated
by each breakdown then stirring the liquid in some way will
maintain a constant quality of liquid in the stressed gap
during a run of test measurements; the electrodes, however,
may change in character during the tests. There may be a
trend in breakdown voltage values giving either 'upward'
conditioning or 'downward' conditioning which can be dis-
cerned from a 'running mean' calculated for say any 10%
sequence out of the total data, as in Fig.2.4. Upward
conditioning, corresponding to a rising trend of voltage,
may be due to the elimination of absorbed gas or of some

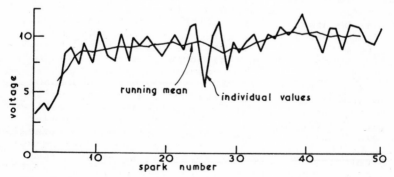

Fig.2.4   Running mean of breakdown voltages

feature on an electrode surface that produces copious electron emission. It may alternatively be due to the carbonisation of fibres or particles attached to an electrode or
even situated in the mid gap. Downward conditioning is most
likely when a relatively large energy is dissipated at breakdown leading to the melting of a greater quantity of electrode metal and the contamination of the liquid by electrode
material and decomposed liquid products especially if these
are not given time to clear from the gap. There is some
evidence that conditioning is polarity dependent such that
if the polarity is reversed or a.c. is used downward conditioning may result, whereas a constant impulse polarity or
d.c. would produce upward conditioning. Upward conditioning
under d.c. may occur without complete breakdown since a
succession of localised discharges can 'clean up' the electrode surfaces and the liquid volumes as well.

2.7   EFFECT OF PARTICLES AND LIQUID MOVEMENT

In contaminated liquids, breakdown processes are domin-
ated by a variety of phenomena associated with the contamination. A wet fibre in an electrically stressed liquid may
eject a stream of water or vapour along which a breakdown
channel may develop. Again, heavy contamination by fibres
may lead to a bridge of fibres lining up between the
electrodes along the direction of the highest field intensity.
This will lead to breakdown at a lower voltage than in the
absence of fibres. Large individual particles may jump
repeatedly from one electrode to the other in a gap stressed
with direct or alternating voltage (2.10).The charge that a
particle acquires when close to one electrode causes it to be
repelled towards the opposing electrode where it gives up its
charge and is promptly repelled back. Using cameras that
take 10,000 frames per second, particles in the region of
100µm size can be observed to cross a 1mm gap in 1 to 10ms.
At the edge of the electrode structure, where the field
fringes, the particles will attempt to follow the curved
paths of the flux lines and may be gradually ejected from
the high field region altogether after several interelectro-
de transits. In other cases particles on the edge of the
field may be drawn into the central high field region.
When a large particle arrives at the electrodes the discharge from it may trigger complete breakdown of the gap.
If oil is pumped through a gap between electrodes, or the
electrodes are rotated, the breaking of fibre chains, movement of gas accumulations and possibly the disturbance to
electrochemical processes at electrodes will affect the
alternating and direct rather than the impulse breakdown
voltage. Typical examples of the effect of particles
and fluid flow with bare electrodes are given in Figs.2.5
and 2.6 (2.11).A dependence of alternating breakdown voltages
on the velocity of flow of mineral oil at ambient temperature was reported by Nelson et al (2.19).An increase of 20%
or more for velocities in the region of only 3mm/s is
ascribed to the prevention of fibre bridges forming
between the bare electrodes used. Charge carriers may also

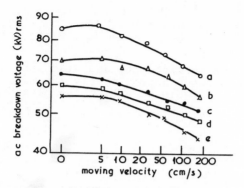

a - filtered   b - 1g of dust   c - 2g of metal dust
d - 5g of metal dust   e - 15g of metal dust

Fig.2.5   Breakdown in contaminated moving oil
          (Ikeda et al (2.11)

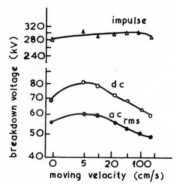

Fig.2.6   Breakdown of moving oil (Ikeda et al(2.11))

be prevented from accumulating near points on the cathode
that give copious electron emission. Lightning impulse
breakdown was not affected by flow. It is unlikely that
liquid flow would have such a large effect or indeed any
effect on conductors covered with cellulose or enamel.
     The effect of particles on conduction and breakdown
is inseparable from that of liquid movement under the action
of an applied electric field. Charged particles, fibres
or clusters of ions will move in the appropriate field
direction carrying liquid with them, a process known as
electroconvection or electrohydrodynamic (EHD) motion. The
liquid will circulate in an electric field as long as there
is a spatial variation in charge density. The steady-state
drift velocity of a single particle of radius R in an
applied field E is

$$\varepsilon\ E^2\ \frac{R}{2\mu} \qquad\qquad \dots\ 2.2$$

where ε and μ are the permittivity and viscosity of the fluid. Fig. 2.7 indicates approximately the maximum velocities attainable in a liquid at ambient temperature

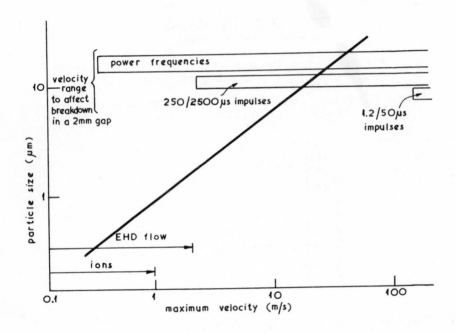

Fig.2.7 Effect of particle movement on breakdown

by particles of different sizes moved by an electric field. It compares these with the velocities of single ions and of fluid motion through EHD processes. For the three forms of voltage cited applied to a 2mm gap, electron movement will be the basic means of charge transfer in all cases, but for 1.2/50µs impulses particles of 100µm size and above alone appear capable of contributing to breakdown by their movement. For switching impulses, characterised by a 250/2500µs waveshape, charge can readily be conveyed by particles in the 5µm to 100µm size range. These are likely to be present in large equipments such as transformers but can be reduced in quantity by repeated filtering through edge-type paper filters or through sintered glass filters. There is a relatively small amount of data on the behaviour of liquid insulation in comparatively uniform fields under switching impulse voltages. Very small particles that it would be impracticable to remove from service equipment, generally those less than 1μm size, are extremely numerous in all liquid insulants. The charge each such particle carries is very small, e.g. 10^{-11}C, and although it is just possible to identify the individual current pulses resulting from the passage of each particle across a gap the charge carried is nowhere near large enough for a single charge to influence breakdown.

2.8 ELECTRON PRODUCTION AND CONDUCTION IN LIQUIDS

Electron emission from a cathode into a liquid subjected to very high applied electric fields can take place by field emission enhanced by geometrical factors such as stress intensification at projecting points possibly aided by incident photons. However, a study of discharge paths in the liquid reveals that the emitted current is very selective as to site and follows one or more filamentary tracks in which relatively high current densities occur.

The apparently special nature of certain cathode sites and their corresponding rarity may explain the observed variations in breakdown voltages in repeat tests.

At lower electric fields, but ones sufficiently high to cause breakdown in contaminated liquids, it is likely that the charge carriers are produced entirely by electrochemical processes at the interface between liquid and metal. A so-called double layer forms at the interface and it has long been known that a highly insulating liquid pumped through a metal tube, or equally between parallel plates without any application of voltage, causes positive ions to be swept into the stream. These accumulate in the bulk fluid where the electrical charge may constitute a hazard through the risk of spark ignition of gases or vapour above the liquid. In the case of liquid insulants, the significance of double layers is that positive ions preferentially, but negative ions also, are created within them. Under the action of a strong electric field at a cathode, electrons can readily be detached from negative ions and indeed can repeatedly revert from negative ions to free electrons and back. This process is assisted by various impurities in the liquid including electrolytes that readily conduct and that create ions.

Once charges originating near the electrodes enter the liquid many conduction processes occur that are not capable of being quantified at present. The mean free path of electrons in a liquid lies in the range 1nm to 10nm whereas in a gas or vapour bubble this may rise to 1μm. Even with an applied voltage gradient of 10^8V/m, which is sustainable only with very pure liquids, and the largest conceivable free path of 10nm the energy gained is 1 eV. The thermal energy of the electrons of 0.025 eV at ambient temperature may be increased perhaps by 0.2eV in one free path and with this energy it may be that electrons can tunnel between trapped positions in the liquid. Electrons can gain enough energy to cause photon emission and there

is also a possibility of ionisation by collision in a gaseous void. The mobilities of free electrons is in the
region of $10^2 cm^2 v^{-1} s^{-1}$ (i.e. 1 mm/ns at 100 kV/mm), depending on the nature of the liquid. Ions are likely to have
mobilities only 10^{-5} times those of free electrons and
various intermediate electronic charge carriers are thought
to exist with about 3 x 10^{-3} times and free electron mobility.
    A discussion of the role of electrochemical processes
and of the electrode/liquid double layer is given by
Felici (2.12).A species in the liquid, probably an impurity,
becomes attached to the cathode where it acquires an
electron to become a negative ion. Under the action of a
high electric field the ion can break away and can also
subsequently release an electron. Meanwhile at the anode,
oxidation of the metal electrode and the presence of the
liquid or its impurities can lead to ion production. The
mobility of these ions is approximately $10^{-3} cm^2 v^{-1} s^{-1}$ so
that even under a voltage gradient of 10kV/cm an ion will
travel only 1 mm in 10ms which is one half cycle at a
frequency of 50 Hz. Hence, conductivity under power frequency a.c. can be expected to be lower than that for d.c.
in a 1 mm gap for stresses below 10kV/cm.
    The variation of steady-state conduction current with
voltage takes the characteristic form of Fig. 2.8. The

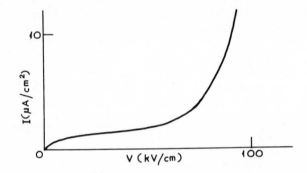

Fig.2.8 D.C. conduction in a liquid

initial part of the curve shows a relatively rapid rise as
charge carriers present in the liquid due to extraneous
radiation, or formed near the electrodes in the double
layers, are progressively removed by the applied voltage.
The more gradual rise that follows is by no means a saturation current but may include additional charge carrier
production as increasing amounts of energy are made available. Accumulating charges that inhibit the rise of
current may also be increasingly cleared by the field and
less ion re-combination may occur as the voltage rises.
The final exponential rise is indicative of charge multiplication through the increasing energies possessed by

moving charge carriers in the liquid, aided by photons.
     The variation of conduction current with time of
voltage application has been described by Sharbaugh et al
(2.13) and is illustrated in Fig. 2.9 for stainless steel

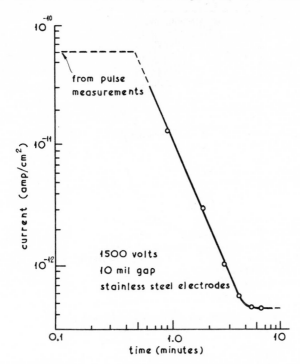

Fig.2.9   Conduction currents in n-hexane
          (Sharbaugh et al (13))

electrodes in n-hexane. For voltage application of more
than a few seconds, a reduction in conduction current is
observed. It is argued by Sharbaugh that the short time
pulse measurements of current are predominantly electrical
phenomena and more meaningful for high voltage insulation
than the steady-state d.c. conduction observed after
several minutes which may have an electrochemical basis.
     Yamashita and Amano (2.14) have measured the variation
of the conduction and prebreakdown currents in filtered
transformer oil following the application of a voltage
step of either polarity of sufficient amplitude to breakdown a 5 mm point/plane gap. The two polarities produce
different forms of current as shown in Fig.2.10 (a) and (b).
     For a positive point there is a delay typically of
3µs while the capacitive current is dying away before a
rapid rise of current of a few mA occurs as shown in
Fig. 2.10 (b). Only 1.0µs after the current rise schlieren
photography shows that a low density region has been
created near the point and this is followed by an exponen-

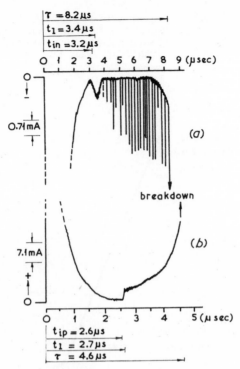

Fig.2.10   Pre-breakdown current from a point
(Yamashita et al (2.14))

tial rise in current leading to breakdown in a further 2 to
3μs; only very small fluctuations are seen on the exponent-
ially rising current. The current corresponding to positive
point breakdown therefore follows a very decisive and rapid
development. With negative point, on the other hand, after
voltage application there is a delay of 3μs again before a
rather slower rise of current occurs to only one fifth the
value for positive point followed by a drop back in current
as shown in Fig. 2.10 (a). There is, however, as before a
density change occuring about 0.1μs after the rise of
current. Then characteristically a series of current
pulses occurs, perhaps twenty of them each of 0.1mA to 1mA
amplitude and just a few ns in duration. During the time
these pulses occur the low density region is advancing in
steps and the charge accumulation is intensifying. After
perhaps 3μs these pulses of current, which are superimposed
on a quite flat background current level, give way to a
sudden exponential rise in current that leads to breakdown.
The negative point gives a formative time lag perhaps twice
as long as that for positive point and is characterised by
lower mean current levels until breakdown is imminent.

2.9   EFFECT OF TEMPERATURE ON THE ELECTRICAL STRENGTH OF LIQUIDS

Most high voltage testing of liquid-insulated equipment takes place soon after manufacture with the insulation cold indeed it could be said that the insulation clearances are designed to meet this requirement. It is a relatively severe test since, even though the insulation may have been thoroughly dried and the liquid degassed, there is a possibility of some gas being trapped, for example in transformer windings. The ability of oil-filled transformers to withstand a test would probably be greater after some weeks of service operation when gas, water and particles have been redistributed throughout the equipment. The mineral oil has to interface with various metal parts, with solid insulation such as cellulose and with gas which may be air or nitrogen. Any equilibrium that may be achieved associated with a constant gas and water content of the oil will be disturbed when variations in loading change the winding temperature. For EHV transformers the use of conservators and breathers that allow only dry air to pass will prevent water ingress to the oil just as a nitrogen 'blanket' does under an expandable diaphragm. For transformers operating at 11kV and below, water ingress is usually not controlled and if a unit is taken out of service for a time it may accumulate free water at the bottom of the tank since cold mineral oil can absorb relatively little water (up to 80ppm). On being energised again the transformer will be as risk if oil circulation picks up some free water.

At the other extreme of temperature, equipment running on emergency overload may become very hot, indeed hot spots in transformers insulated with mineral oil and cellulose may reach 180°C, as discussed by Long (2.15). Faced with the prospect of disconnecting consumers or shutting down a critical plant such as a nuclear reactor following a plant outage, system operators may prefer to load a remaining transformer to 200% for a few minutes. The reduction in insulation life through decline in the mechanical and electrical properties caused by temporary overloads is relatively well understood. However, less information is available on a.c. and impulse strength of oil-immersed insulation at elevated temperature. Gas and water evolved by chemical action in oil, in cellulose and at the conductor surface (2.16) can be absorbed more readily in hot mineral oil until the point of saturation is reached. At some temperature the water absorbed in the oil may vaporise to form bubbles and the lower the water content before an overload the better it can be withstood. The oil will usually be saturated with air or nitrogen at some temperature at which it has operated in the previous period. A sudden temperature rise will allow it to absorb more gas but if the oil dissociates at high temperature bubbles may still form. The insulation may also be at risk as it cools after an overload since a heavy gas absorbtion in the oil will then be released as bubbles.

Apart from the dynamic equilibrium of the gas and water content of the oil the variation of several parameters with

temperature must be considered, particularly viscosity,
surface tension and density. The viscosity and density
affect the rate of flow of the insulant through ducts or
past high voltage parts, especially for insulation cooled
by natural convection such as in distribution transformers,
but units with forced cooling are also affected. The rate
of flow of liquid does affect electrical strength, assist-
ing it through the breaking of fibre bridges and the dis-
persal of developing discharges but reducing it if streams
of bubbles are carried into highly stressed regions as
might happen at very high temperature. As viscosity falls,
the electroconvection, or EHD processes, which can assist
breakdown will be facilitated. It is characteristic of the
surface tension of liquids to fall with increase of temperature and this affects bubble formation. Bubbles will then
grow larger for a given discharge energy and the pressure
inside the bubble will be lower with the bubble more likely
to elongate in the direction of the field or join up with
other bubbles. Also bubbles of a smaller size will become
viable. With lower gas or vapour pressure in a bubble the
Townsend ionisation coefficient will be larger under a given
applied electric stress while an avalanche proceeding across
a larger bubble diameter will be more likely to result in
streamer development. Bubbles in oil are typically 50μm or
more in diameter and the internal pressure ranges up to 4
bars but it is said that vibration can release much smaller
bubbles from cellulose (2.16).

The characteristic variation with temperature of the
electrical breakdown voltage of a liquid-insulated equipment is shown in Fig. 2.11, as given by Lewis (2.17). Over
the more usual operating temperature range for liquid

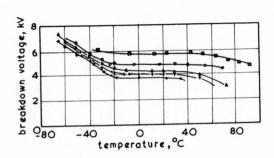

n-decane (highest)   n-octane   n-heptane   n-hexane
n-pentane

Fig.2.11   Effect of temperature in breakdown voltage
(Lewis (17))

insulation, such as mineral oil, the breakdown voltage may
rise slightly between 20ºC and 50ºC and then fall, usually
with a steady decline above 100ºC. This will depend on the
characteristics of all the materials used in the insulation

structure and on their gas and water contents.

To illustrate the theoretical approach to breakdown and its prediction of the effect of temperature it is useful to take a model such as that put forward by Kao (2.18) based on instability arising in a filamentary current path. An energy balance is considered for a current filament during application of a step voltage of short duration. The critical applied voltage gradient Eav needed to produce breakdown is related to the initial temperature T_oK of the liquid and its boiling point T_bK by

$$Eav = \frac{8}{9} \frac{Cpgkd}{W\varepsilon\mu_oMI} \left[To^2 \exp\left(\frac{W}{kTo}\right) - Tb^2 \exp\left(\frac{W}{kTb}\right) \right]^{\frac{1}{3}} \quad \dots \ 2.3$$

The charge carrier mobility is $\mu o \exp\left(-\frac{W}{kT}\right)$, where k is Boltzmann's constant, W is an activation energy and μo is the mobility factor. The mean values of the relevant parameters of the liquid between To and Tb are specific heat, Cp, density g and absolute permittivity ε while d is the gap length and I the duration of time the voltage has been applied. A geometric factor M that lies between 0 and 1 depends on r/d, where r is the radius of the filament, and is given by Kao (18). This model can be used to predict the effect of pressure and temperature on impulse breakdown of a liquid for which data is available.

The effect of flow of gas-saturated mineral oil at temperature up to 80°C was shown by Nelson et al (2.19) to be a sharp fall in alternating breakdown voltage. This was attributed to bubbles in the oil being carried into regions of high stress and indeed the oil temperature may have fallen as it moved round the circuit so encouraging bubble formation. In using bulk liquid insulants that are pumped round a transformer and radiators, gas evolved in some part of the circuit from where it could move into highly stressed areas of insulation can be a hazard.

REFERENCES

2.1.   B.S. 148 1972 Insulating oils for transformers and switchgear.

2.2.   IEC 296 1969 Specification for new insulating oils for transformers and switchgear.

2.3.   Moore, C.L. and Mitchell, G.F. Design and performance characteristics of gas/vapour transformers, 1982, IEEE Trans., PAS-101, 2167-2170.

2.4.   Allan, R.N. and Hizal, F.M. Prebreakdown phenomena in transformer oil subjected to non-uniform fields, 1974, Proc. IEE, 121, 227-231.

2.5.   Chadband, W.G. and Calderwood, J.H. The propagation of discharges in dielectric liquids, 1979. J Electro statics 7, 75-91.

2.6.   McGrath, P.B. and Nelson, J.K. Optical studies of prebreakdown events in liquid dielectrics, 1977, Proc IEE, 124, 183-187.

2.7.   Wong, P.O. and Forster, E.O. The dynamics of electrical breakdown in liquid hydrocarbons, 1982 IEE Trans El-17, 203-220.

2.8.   Forster, E.O. The search for universal features of electrical breakdown in solids, liquids and gases, 1982, IEEE Trans El-17, 517-521.

2.9.   Schmidt, W.F. Elementary processes in the development of the electrical breakdown of liquids, 1982, IEEE Trans El-17, 478-483.

2.10.   Molinari, G and Viviani, A, Analysis of the charge exchange mechanisms between impurities and electrodes in a dielectric liquid, 1979, J. Electrostatics 7, 27-32.

2.11.   Ikeda, M., Teranishi, T, Honda, M and Yanari, T. Breakdown characteristics of moving transformer oil, 1981, IEEE Trans, PAS-100, 921-928.

2.12.   Felici, N.J., A tentative explanation of the voltage - current characteristic of dielectric liquids, 1982, J Electrostatics, 12, 165-172.

2.13.   Sharbough, A.H. Devins, J.C., and Rzad, S.J. Progress in the field of electric breakdown in dielectric liquids, 1978, IEEE Trans. El-13, 249-276.

2.14.   Yamashita, H and Amano, H, Prebreakdown density change current and light emission in transformer oil under non-uniform field, 1982, J Electrostatics 12, 253-263.

2.15.   Long, L.W. Overloading and loading limitations of large transformers, 1983, Electra, 86, 33-51.

2.16.   Heinrichs, F.W., Bubble formation in power transformer windings at overload temperatures, 1979, IEEE Trans, PAS-98, 1576-1582.

2.17.   Lewis, J.J. Electric breakdown in organic liquids, 1953, JIEE, 100, 11A, 141-148.

2.18.   Kao, K.C. Theory of high-field electric conduction and breakdown in dielectric liquids, 1976, IEEE Trans El-11,121-128.

2.19.   Nelson, J.K., Salvage, G and Sharpley, W.A., Electric strength of transformer oil for large electrode areas, 1971, Proc IEE, 118, 388-393.

Chapter 3

Breakdown in solids

R. Cooper

3.1 INTRODUCTION

Practical work has not only demonstrated the inadequacy of theories but also the existence of many secondary processes which interact with, and even supersede the electronic one in causing breakdown. Besides the primary electron generation and removal processes the following must be considered:

(1) the heating effect of pre-breakdown current and dielectric losses;

(2) the compressive force due to charges on the dielectric faces;

(3) "treeing", which may be caused by
(a) discharges in the ambient at electrode edges,
(b) discharges in gas filled cavities (voids) within the dielectric-
(c) stress concentration at small conducting inclusions, and sharp electrode edges;

(4) electrochemical reactions with the environment, and also following the dissociation of impurities, which may cause loss of insulating properties.

In well designed, carefully executed experiments in which the stressed volume, and the duration of the stress are minimal, and the ambient temperature is not too great, the above effects can be avoided. Electric strengths (F_b), (given by the quotient of the breakdown voltage (V_b) and the thickness (d) of a planar specimen), of about $1MV.cm^{-1}$, and about $10MV\ cm^{-1}$ can be achieved with crystalline substances and polymers respectively. In practice the mean working stress seldom exceeds $150\ kV.cm^{-1}$, it may be only $50kV.cm^{-1}$ in plastics, and less in a stator winding.

Electric strength determined in the above special circumstances is called "intrinsic", this word reflecting history in that early theories stimulated belief in the existence of a physical constant, independent of geometry, electrode material, voltage waveform, and determined by molecular and/or crystal structure and temperature. The concept, is not supported by experiment.

3.2   SOURCES OF CURRENT CARRIERS

The conductivity is not negligible when the field (F) exceeds about 10^5V.cm^{-1}, and it increases rapidly with (F). Possible sources (3.1) of conduction electrons are:

(a) Fowler-Nordheim (field) emission yielding cathode electron current density

$$j_c = CF^2 \exp\left(\frac{-D}{F}\right) \qquad\qquad 3.1$$

(b) field aided thermionic emmision (Schottky) effect), yielding at the cathode, a current density

$$j_s = AT^2 \exp-\frac{(\phi - BF^{\frac{1}{2}})}{kT} \qquad\qquad 3.2$$

(c) field enhanced ionisation of impurities (Poole-Frenkel effect).

In the above, A,B and D are constants, ϕ the energy difference between the Fermi level of the metal and the bottom of the conduction band, $C \propto 1/\phi^{\frac{3}{2}}$, k is Boltzmann's constant, and T is temperature (^{O}K). The Poole-Frenkel effect yields a conductivity proportional to $\exp(F^{\frac{1}{2}}/kT)$; the origin of the field dependence is the effect of electric intensity on the height of the potential barrier which must be surmounted by the escaping electron, and the explanation is the same for the Schottky effect. The electron is transported to the anode either by "hopping" from one centre to an adjacent one, or by drifting from trap to trap with a velocity in the field direction superposed on random thermal motion. The average drift distance is proportional to the applied field and electron mobility, and inversely to the trap density, which is related to the density of impurities and imperfections. If the specimen is thin and F is great, and electron is likely to cross without trapping, but with thicker specimens and/or weaker field, the space charge of trapped electrons will modify the assumed uniform field. Trap density depends on structure, but with good alkali halide single crystals about 0.5mm thick at 20°C or less, an effect is unlikely (3.2) if $F > 10^5 \text{V.cm}^{-1}$. However, trapped injected charge in polymers only 0.05mm, thick can be demonstrated by the method of Thermally Stimulated Currents (3.3). The effect of such charge on the electric strength $(Vb/_d)$ of a number of polymers has been demonstrated (3.4, 3.5, 3.6), by using combinations of direct and superposed voltages. Fig. 3.1 illustrates the behaviour of polyethylene about 50μm thick at 20°C.

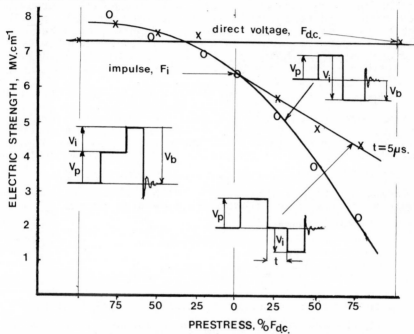

Fig. 3.1   Effect of prestress on average electric strength of Alkathene 7.

The space charge apparently forms in about 10^{-4}S and decreases to a negligible amount in about 10^{-4}S, when the specimen is short circuited. Injected space charge is a factor, but certainly not the only one, having bearing on the frequently reported differences between electric strength measured with impulse, direct, and alternating voltage.

3.3   THERMAL BREAKDOWN

Many substances also conduct by transport of ions. An ion moves from one place of equilibrium to another by acquiring energy (W) to surmount the potential hill between. Hence the ionic conductivity (σ) is

$$\sigma = \sigma_o \exp \left(\frac{-W}{kT} \right) \qquad\qquad 3.3$$

and this strongly temperature dependent component is added
to the electronic component. Besides energy loss due to
conduction, there is the dielectric loss due to dipoles which
varies with temperature according to the loss factor, tan δ.
This may be represented by an effective conductivity,
$\varepsilon_o \varepsilon_r \omega$ tan δ which increases with T only in certain ranges.
Heating is also caused by discharges in voids and by ambient
discharges. Very likely the rate of heating will increase
with increasing temperature. The heat generated is partly
lost by conduction to the surface, and partly absorbed. The
rise in temperature enhances the electrical conductivity and
so exacerbates the heating. Thermal instability (3.8,3.9)
is possible. The energy balance equation is

$$\sigma_{(T,F,)} \cdot F^2 = C_V \ (dT/dt) \ - \ \text{div (K grad T)}; \quad 3.4$$

C_V and K are respectively the specific heat and thermal
conductivity, not necessarily independent of T. A general
solution is not available, but the special case has been
considered of a slab of thickness (d), large area,
conductivity $\sigma = \sigma_o \ \exp \ (\frac{-W}{kT})$, and with electrodes
providing no thermal impedance. If the voltage is increased
slowly so that a quasi-steady state obtains $C_v \ (dT/_{dt}) = 0$.
The breakdown voltage

$$V_{bm} \ = \ \left(\frac{8K \ kT_o^2}{\sigma_o W} \right)^{\frac{1}{2}} \ x \ \exp \ \frac{-W}{2kT_o} \quad 3.5$$

T_o is the ambient temperature; usually $W > kT_o$ and the
"maximum thermal voltage" (V_{bm}) does not depend on any
temperature at which physical or chemical change occurs but
only on the ambient temperature (T_o). Normally $W > kT_o$ and
V_{bm} decreases with increase in T_o because the exponential
term dominates the bracketed one; V_{bm}, does not depend on d
and $F_b \propto \frac{1}{d}$. With impulses time does not permit much heat
loss by conduction and therefore div (K grad T) = 0. The
"thermal impulse breakdown voltage" V_{bi} is obtained:

$$V_{bi} \ = \ \left(\frac{3C_v \ kT_o^2}{\sigma_o W \ t_b} \right)^{\frac{1}{2}} \ \exp \ \frac{W}{2kT_o} \quad 3.6$$

t_b is the time to breakdown and the applied voltage (V_t) is
assumed to increase linearly with t, i.e. $V_t = V_{bi} \ (^t/tb)$.
As in the previous case V_{bi} decreases with increase in
ambient temperature. Other theories of breakdown predict
this but thermal breakdown voltage is proportional to

(specimen resistance)$^{\frac{1}{2}}$, because of the factor $\frac{1}{\sigma_o}$ exp $^W/2kT_o$. A further result is that with constant field F_m, the temperature at the hottest point increases with t asymptotically to the maximum T_m. When $F < F_m$, T reaches T_m in finite time and continues to increase continuously. T_m is a critical value, not related to any physical or chemical change.

3.4   ELECTRONIC THEORIES

3.4.1 Theories of Frohlich

Probably, the most successful and certainly the most stimulating of the early theories are those of Frohlich (3.1) (3.9). The predicted electric strengths are truly intrinsic. Neither specimen size nor electrode dependent parameters are involved in predicted electric strengths. The 'high temperature' or 'amorphous' theory (3.10) assumes an energy band model consisting of the conduction band separated from the valency band by several eV. At energy 2V (one or two eV) below the conduction band the ground level of an impurity exists and its excited levels extend effectively as a continuum over 2ΔV below the bottom of the conduction band. The dielectric temperature (T), and 2V, are such that there are many excited and conduction electrons. It is assumed that interactions between conduction electrons and the dielectric structure are much less likely than interactions between conduction and excited electrons, and the conduction electrons themselves. The conduction electrons gain energy from the field and this energy is shared amongst all electrons because of the interactions. Some of the excited electrons are raised to the conduction band, but stability demands that others return from the conduction band. In becoming trapped these give up energy, which excites structural vibration. In effect the effective electron temperature (T_e) exceeds the structure temperature T_o, permitting the energy transfer.

There is a critical field

$$F_c = G \exp (\Delta V/2k \ T_o) \qquad\qquad 3.7$$

where G is a constant independent of T_o, above which the mechanism of exchange no longer can establish a balance. The electron temperature 'runs away' and the dielectric is destroyed. No thermal constants are involved, and only the electronic band structure. However, the assumption of energy exchange by electron-electron interactions cannot be justified at all temperatures because the electron densities decrease with temperature. The "low temperature" or

"crystalline" theory applies when the energy loss is
predominantly by direct excitation of lattice vibrations and
it was developed specifically for the alkali halides.
Consider conduction electrons with mean energy (E). Each
gains energy from the field for average time $T(E)$ and then
looses it in exciting lattice vibrations. The rate of gain
of energy A, (F,E,T,) is proportional of F^2 and $T(E)$. For
stability A(F.E.T.) must equal the average rate of loss to
the lattice B (E,T). At each interaction the electron either
emits or absorbs vibrational quanta $h\delta$; h is Planck's
constant and δ is the vibrational frequency. If there is
only one frequency, the rate of loss of energy is the
product of $h\delta$, the number of interactions/sec. which
increases slowly with temperature, and the fraction giving
energy loss. Each alkali halide exibits a strong optical
vibration; the assumption of a single frequency is justified
and A(F.E.T.) and B(E.T.) can be calculated with results
illustrated by Fig. 3.2.

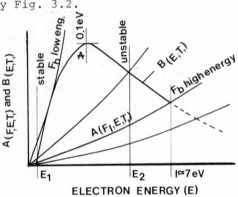

Fig. 3.2   Average rate of energy gain A(F.E.T.) from
            applied field (F), and average rate of
            energy loss to lattice B(E.T.).

There are two possible breakdown criteria

(a)   von Hippel's, or "low energy" (3.12) criterion,
      when the field must accelerate all electrons.

(b)   Frohlich's "high energy" criterion requiring the
      field to accelerate only electrons with ionizing
      energy. (3.11)

In support of (b) Frohlich points out that (a) ignores
the random fluctuations in the energy of an electron, which
render impossible a steady state in any field, however weak.
The proposed stability criterion is that the rate of electron
loss by recombination is equal to the rate of electron
creation by collision ionization. In the latter a fast
electron looses energy and two slow electrons are produced.
Recombination occurs following a collision between two
electrons in which one, at the expense of the other, gains
enough energy to exceed I, thus replacing a lost fast
electron. Frohlich proposed that F_I is the greatest field in

which equilibrium is possible. This implies anisotropy of
electric strength because breakdown is determined by the
properties of electrons with energy 5 to 10 eV, or more in
contrast with the alternative where electrons of thermal
energy, and therefore those insensitive to the lattice
structure, are dominant. There is also implied a transition
from "low temperature" to "high temperature" behaviour
indicated by change in the gradient of F_b plotted against T
from small, (and positive for alkali halides) at low
temperatures to appreciable and negative at high temperatures.
Frohlich's theories are compatible with a number of facts.

    (a)   the alkali halides are anistropic

    (b)   the calculated electric strengths agree as well
            as can be expected with measured values.

    (c)   imperfections will increase F_b because they
            reduce the electron free path. There is
            evidence of this with alkali halides: there is
            some evidence with polymers; introduction of
            electron scattering, dipolar elements increases
            F_b in the "low temperature" region.

However, there are also facts which cannot easily be
accommodated; viz:

    (d) F_b has been found to decrease with d for a
            number of substances. A compilation (3.15) of
            results for NaCl shows the decrease is by a
            factor of 10 over three decades increase of d.

    (e)   breakdown characteristics typical of a number
            of substances are illustrated by Fig.3.3.(over)
            Clearly, the "high temperature" law (Eqn.3.7)
            is not obeyed.

    (f)   the prebreakdown luminescence, attributed to
            excitation of centres by the rapidly growing
            number of energetic electrons, or by the
            radiation from direct recombination of
            conduction electrons and holes, propagates
            from cathode to anode (3.16), even when the
            former is the plane of an extremely
            assymmetrical point/plane system (3.17). This
            is illustrated by Fig 3.4

            Clearly, the criterion, rate of ionization $\geq$
            rate of recombination is not by itself complete.
            Cathode phenomena must be included in a
            satisfactory explanation of breakdown.

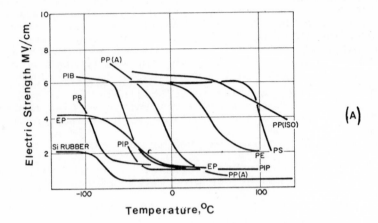

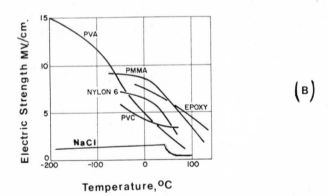

Fig. 3.3   Electric strength of (A) non-polar polymers
and (B) polar polymers, and an ionic crystal
NaC1. From compilation by R.A. Fava, 1977,
'Electric breakdown in polymers', Treatise
on Materials Science and Technology, Vol.10B
pp. 677-739 Academic Press, New York.
PE=low density polyethylene PP=polypropy-
lene (atactic and isotactic); PIB= poly-
isobutylene EP=ethylene-propylene copolymer,
PIP=polyisoprene; PMAA=polymethyl methacrylate; PVC=Polyvinyl chloride;PA=polyvinyl
alcohol.

3.4.2. Current Continuity, and Runaway

Interpretation of Fig.3.4 requires a mechanism of stress transfer because the substance will luminesce only when the field is sufficient to ensure the necessary energy transfer to the luminescent centres. Space charge distortion of the initial field distribution is indicated. This occurs if electrons move to the anode more rapidly than the positive holes do to the cathode. The resulting positive space charge enhances the Fowler-Nordheim cathode current (j_c) and also the progeny of collision ionisation because the numbers are proportional to j_c.

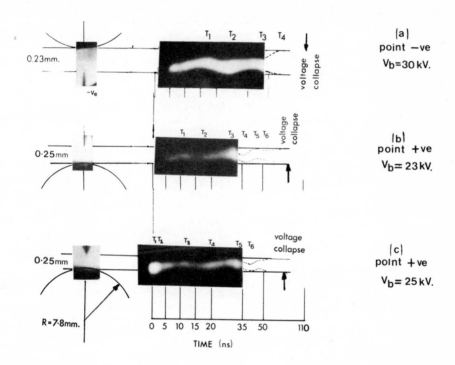

Fig. 3.4   Streak photographs of pre-breakdown luminescence demonstrate the importance of cathode conditions. In (C) the streak trigger was delayed, causing light to be integrated thus producing the spot at time zero.

The growth of current carrier density is opposed by:

(a)   removal of carriers at the electrodes:

(b)   recombination of electrons and holes: The ionization coefficient (α) is assumed proportional to exp ($-H/F$) where H is a constant, Poisson's equation relates F to the nett space charge density, and the rate of recombination is proportional to the density of electrons (N_-) and the density of holes (N_+). No general solution is available.

Cooper and Elliott (3.16), referring specifically to Kbr, assumed hole mobility to be negligible, the current density equal to j_c throughout, and ionization balanced by recombination. It was deduced from considering equilibrium that nett positive space charge would accumulate if $\alpha\mu e/$(constant) exceeded $j_c/\mu E$, where μ is electron mobility. They observed that breakdown is preceded by a time lag, the mean value depending on the cathode, and the number/time distribution indicating a chance event. The Fowler-Nordheim current j_c is 'noisy' and α is only a mean. It was considered that instability resulted from a fluctuation in one of the parameters. The cathode field determines j_c and it depends on the total positive space charge, which increases with d. Thus F_b may be expected to decrease with increasing d. In contrast with the above, O'Dwyer (3.15, 3.18) has assumed that recombination can be neglected and the holes have small mobility. Klein and Solomon (3.19), have considered both the case when carrier removal is by recombination, and when it is by drift to the electrodes. If reasonable values are assumed for parameters, current runaway is predicted when the cathode field exceeds the mean field by a few percent and regarding the thickness effect, O'Dwyer (3.15) has obtained predicted values for NaC1 which agree well with the measured values over the range 5×10^{-3}mm. to 1mm.

3.4.3. Avalanche Breakdown

Seitz (3.20) has proposed that breakdown will ensue from the passage of a single avalanche, provide the energy transferred by the electrons to the tubule through which the avalanche passed amounted to about 10eV per atom. The avalanche size depends on specimen thickness because the tubule radius $r = (2Dt)^{\frac{1}{2}}$ where D is the diffusion coefficient, and the electron transit time, $t \propto d$. Consequently, the theory predicts a slow decrease in F_b with increase in d, which in magnitude agrees well with experiment. However, the necessary avalanche would contain 10^{11}-10^{12} electrons passing through a tubule of radius about 10^{-3}cm. at most.

The electron density is such that fields exceeding $10^{10}Vcm^{-1}$ would exist about the avalanche head. Such an avalanche is unlikely because its development would be quenched at an early stage by;

(a)   attraction of escaping electrons back to the avalanche head.

(b)   reduction of the field between the avalanche head and the anode by the space charge.

Klein (3.21) has discussed this with reference to the breakdown of thin film, i.e. 500A - 30,000A thick) planar capacitors. When the electrodes are very thin the current arising from the discharge of self capacitance into the developing channel causes the electrode to evaporate locally, and isolate the breakdown region. Very many breakdowns can occur with the same specimen. Apparently they occur at random over the electrode area, and each breakdown affects only about $10^{-8}- 10^{-9}cm^2$. This, along with the observation of statistical time lags, and the thickness dependance of F_b has encouraged interpretation in terms of avalanche formation. It is considered that breakdown results from the passage of many self quenched avalanches, all originating at the same cathode spot. The positive charges of the first avalanche drift to the cathode, enhance the field locally and increase the chance of a second avalanche from the spot.

Successive avalanches sustain the growth of the positive space charge, the localized enhancement of cathode field, the avalanche rate, leading to current instability along the avalanche path.

3.5   POLYMERS AT HIGH TEMPERATURES

3.5.1. Electro-mechanical breakdown ✓

The rapid decrease of F_b exhibited by many of the characteristics of Fig. 3.3 occurs at temperatures above the softening point. This promotes speculation about a connection between F_b and mechanical properties which is encouraged further by the experiments for example of Pais (3.22) which revealed a direct proportionality between the hardness of thermosetting allyl polymers cured to different extents and F_b. It was suggested that both electrical and mechanical response originate from molecular motions and although different parts of the molecular chain may be involved, it follows that the time/temperature variations are similar. This is true for electrical and mechanical loss characteristics. A compression of $(\tfrac{1}{2}\varepsilon_o \varepsilon_r F^2)N.m^2$ is experienced, and this is appreciable when $F > 10^5Vm-1$. This force is restrained by the elastic force but the stress-strain relationship is usually not simple for a polymer, or unknown. Stark and Garton (3.23) assumed a logarithmic law for polyethylene and for equilibrium

$$\tfrac{1}{2} \, \varepsilon_o \, \varepsilon_r \, F = Y \log \frac{d_o}{d} \qquad\qquad 3.8$$

where Y is Young's modulus, d_o the initial thickness. A stable state is impossible when $\frac{d}{d_o} < 0.6$ because $d^2 \log(d_o/d)$ reaches its maximum when $d/d_o = \exp(-\tfrac{1}{2}) = 0.6$. If this compression has occurred before the electric strength has been achieved the specimen will collapse. The critical electric stress is $(Y/\varepsilon_r \varepsilon_o)^{\tfrac{1}{2}}$ and the greatest value of F_b that can be observed is

$$V_b/d_o = (Y/\varepsilon_o \varepsilon_r)^{\tfrac{1}{2}} \times 0.6 \qquad\qquad 3.9$$

Electromechanical breakdown occurs with polyethylene in the "high temperature" region and the characteristic has been shown to follow changes in elasticity induced by radiation cross-linking. Fava (3.24) has investigated the effect using an optical level and Blok and LeGrand (3.25) have used photo-elasticity. Apparently electromechanical breakdown of polyethylene can be prevented by encapsulation (3.25, 3.26, 3.27) in a glassy substance and the "high temperature" characteristic is suppressed by the constraint so provided.

3.5.2.   Electric strength and Molecular mobility

The electromechanical hypothesis does not explain all the observations with polymers in the "high temperature" region. For example, with polymethylmethacrylate F_b starts to decrease at about 20^oC, when the dielectric is glass hard. Dynamic mechanical losses occur in all polymers and vary due to the onset at various temperatures of mobility in different parts of the molecular chain. This mobility is accompanied by an increase in "free volume" between molecular chains, and a sudden change in the specific heat and coefficient of thermal expansion. There is also evidence of corresponding abrupt changes in the activation energy for electrical conduction. Sabuni and Nelson (3.22) have investigated the breakdown of polybutadiene, polystyrene, a 75/25 copolymer, and a 52/48 copolymer in the range -100^oC to 100^oC. The glass-rubber transition temperatures determined by differential thermal analysis occurred respectively at -93^oC, $+85^oC$, -83^oC and 25^oC., corresponding with the temperatures at which the electric strengths decreased, step like. Artbauer assumed the structural model of Hirai and Eyring (3.29), in which free volume is assumed accommodated as holes in the polymer structure which are being created and filled in at random by the molecular motion. The chance clustering of holes is assumed to increase the probability of long electron free paths, and hence the rate of ionization. Artbauer (3.30) deduced F_b = constant/(electron mean free path), the temperature dependence simply following the temperature variation of the denominator. A further result is that F_b is dependant on the duration of voltage application (t) and when t is smaller than the molecular relaxation time, F_b is constant. Relaxation processes are discussed fully by McCrum (3.31); polymethylmethacrylate although glass hard at 20ºC exhibits a transition attributed to the hindered rotation of the alkyl side group- $COOCH_3$ in the molecule. At about 20°C., F_b experiences an appreciable increase in the negative temperature gradient.

3.6   DIVERGING FIELDS AND "TREEING"

Many investigations (3.32, 3.33, 3.34), have been made with point-plane arrangement because it is a convenient way of initiating the industrially important "treeing" phenomenon. In the absence of space charge the stress at a point, radius r, is approximately

$$F_m = (2V/r) \times \log (1 + \frac{4d}{r})   \qquad 3.10$$

Calculation usually yields F_m at breakdown appreciably greater than the electric strength of the same substance between plane electrodes at the same separation (d). Also V_b with the point negative is appreciably greater than with the same arrangement and point positive. Interpretations usually follow Mason's (3.32) hypothesis that breakdown occurs when the "intrinsic strength" is exceeded over the small volume next to the point. This is rendered conducting with the result that the high field region and the breakdown channel propagates to the plane electrode. The failure of the above formula is attributed to:

- a)   the effect of stress dependent conductivity
- b)   the space charge at the point which when negative exceeds that for the opposite polarity

In effect, the point is blunted. A recent investigation (3.17), using streak photography has revealed, in the case of cast polyester resin specimens with $(\frac{d}{r})$ <100, and d<0.5mm. that collapse of voltage was preceded by luminescence propagating from cathode to anode with velocity V $\simeq 10^6$ cm.s^{-1} both when the cathode was the plane electrode, and the point and just as with planar specimens, the observations are accounted for by the establishment of a critical cathode field through pre-breakdown conduction. Voltage collapse occurred some tens of ns after the glow had filled the inter-electrode space, but destruction of dielectric was restricted to a very narrow tubule along the axis in the centre of the glow i.e. along the probable line of maximum

current density. Single frame shuttered camera (exposure
$\cong$10ns) photographs are consistent with the above for
polyester resin and solid argon. However, when $(d/r) > 100$,
d > 0.5mm. another process intervenes. With application of
a succession of voltage impulses, a tree like structure of
micro-cracks spreads from the point and as it extends to the
plane the micro-cracks broaden and eventually support
discharges. The growth of the tree is illustrated by Fig.3.5
(over) and breakdown eventually occurs sometime after the
pattern of cracks/tubules has joined the electrodes and when
there is a path through it which is broad enough to permit
development of a spark (3.37). The micro cracks are
electrical in origin, for they are associated with the
appearance of a 20ns pulse of light originating principally
near to the point and also as a band across the specimen, in
front of the plane. They are attributed to shock waves
resulting from vapourization of a tiny region at the point.
The luminosity is to be associated with incipient current
runaway. It is not clear why the process is quenched when
$(d/r > 100$, but charge carrier transit in such specimens
must be hindered by trapping in the extensive region of weak
field in front of the plane electrode. With respect to
crack formation, it is appropriate to note that Budenstein
(3.38) has described a number of electro-chemical reactions
which can take place when the charge density is sufficient.
These do not depend on ionization by collision but lead to
creation of free electrons and the rapid production of gas.
Budenstein considers the first stage of breakdown to be the
production by such reactions of a tubule of gas and it is
through this that breakdown occurs.

'Treeing' also readily occurs at relatively low
voltages provided the stress is alternating. The "tree"
develops from points of stress concentration e.g. a sharp
electrode edge, a small conducting inclusion, a point where
electrical discharges in a weaker medium impinge on the
dielectric surface. A gas filled cavity within the
dielectric provides an example. The stress on the gas in
the cavity is enhanced by a factor depending on the relative
permittivity of the solid (ε_r), and the geometry, and it is

ε_r for a flat, thin disc when the field is perpendicular to

its ends. Discharges in cavities are discussed in detail in
Chapter 4. The breakdown strength of air gaps, length $\cong$ 1mm.
at NTP is about 5kV./mm. A spark in a gas requires;

     a)   replacement of the initiating electron.
     b)   maintenance of the field so that ionization
         rate is maintained.

Each discharge in a void does not produce a spark because
the charges deposited on the dielectric surface "back off"
the applied field. But the surface is under continued
bombardment alternately with positive and negative charges.
Electrons can penetrate some distance into the dielectric
despite a surface layer of + ions. The surface is eroded
and eventually a tree grows.

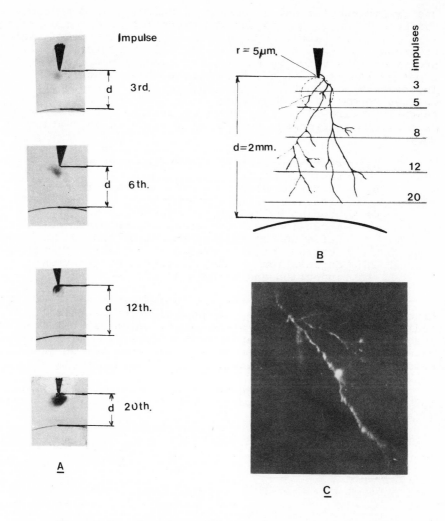

Fig. 3.5   Treeing with impulse voltages. With
successive applications of the inception
impulse the tree advances from the point
to the plain. (B) but the luminous
region (A) and hence the high field
region remains at the point. (C
illustrates the faceted appearance of
the damage at X 400 magnification with
dark field illumination.

There are two stages in breakdown by a.c.treeing. First,
the inception period which may be for very many cycles. This
is followed by a relatively short period of tree growth.
Discharge detection reveals that the transition is
accompanied by a big increase in the discharge magnitude,
which persists until breakdown. The tree consists of hollow

gas filled tubules which sustain discharges. A discharge has been likened to a conducting extension of the electrode (3.32, 3.33). Extension is assumed by "localized intrinsic breakdown" at the tip. This is unlikely because it implies preferential advance by the leading tubule; many have reported random growths. Moreover, although tubules luminesce, the length is not equipotential, because of charges deposited from growing avalanches. Conditions favouring spark formation cannot be assumed. Several hypothesis have been proposed to account for growth initiation . They are;

a) thermal decomposition in the high stress region.
b) pulsating electromechanical pressure producing fatigue cracking
c) erosion by undetected discharges in air inclusions due to incomplete wetting, or differential thermal expansion and contraction at surfaces of contact
d) degradation of the dielectric by recurrent injection and withdrawal of electrons.

There is evidence that in laboratories at least, treeing in polyethylene can be controlled by organic additives (called voltage stabilizers) which are believed to absorb energy from, or trap injected electrons, or in the case of voids, create a conducting film over the surface which short circuits the void.

More recently, treeing due to the presence of water and sulphides has been reported and these are discussed later (Chapter 15).

REFERENCES

3.1   O'Dwyer, J.J., 1973, "The theory of electrical
        conduction and breakdown in solid dielectrics",
        Clarendon Press, Oxford.

3.2   Cooper, R., 1962, "The electrical breakdown of alkali
        halide crystals" Progess in Dielectrics, Vol.5,
        pp. 97-141, Heywood and Co., Limited, London.

3.3   Cresswell, R.A. and Perlman, M.M., 1970, "Thermal
        currents from corona charged mylar"., J. Appl. Phys.,
        41, pp. 2365-2369.

3.4   Bradwell, A., Cooper, R. and Varlow, B.R., 1971,
        "Conduction in polythene with strong electric fields
        and the effect of prestressing on the electric
        strength"., Proc. I.E.E., 118, pp. 247-254.

3.5   Cooper, R., Varlow, B.R. and White, J.P., 1975,
        "The effect of prestressing on the electric strength
        of some polymers"., I.E.E. Conf. Publ., 129, pp.209-
        212.

3.6   Kitani, I. and Arii, K., 1982, "Impulse breakdown of
        prestressed polyethylene films in the ns range"
        I.E.E.E. Trans. EI-17, No.6., pp.571-576.

3.7   O'Dwyer, J.J., 1973, ibid. p.18.

3.8   O'Dwyer, J.J., 1973, ibid., Chapter 6.

3.9   Whitehead, S., 1951, "Dielectric breakdown of solids",
        Chapter 3, Clarendon Press, Oxford.

3.10   Frohlich, H., 1947, "On the theory of dielectric
        breakdown in solids", Proc. Royal Soc., A.188,
        pp. 521-528.

3.11   Frohlich, H., 1939, "Dielectric breakdown in solids"
        Reports on progress in Physics, 6, pp.411-430,
        Physical Society, London.

3.12   Callen, H.B., 1949, "Electric breakdown in ionic
        crystals", Phys. Rev., 76, pp. 1394-2001.

3.13   Cooper, R. and Fernandez, A., 1961, "Anisotrophy in
        the electric strength of alkali halide crystals",
        J. Phys. (D), 18, pp. 1262-1263.

3.14   Cooper, R. and Elliott, C.T., 1968, "Directional
        electric breakdown of KC1 single crystals",
        J. Phys. (D), Series 2, Vol. 1, pp. 121-124.

3.15   O'Dwyer, J.J., 1967, "The theory of avalanche break-
        down in solid dielectrics", J. Phys. Chem. Solids,
        28, pp. 1137-1144.

3.16   Cooper, R. and Elliott, C.T., 1966, "Formative
        processes in the electric breakdown of potassium
        bromide", J. Phys. (D), 17, pp.481-488.

3.17   Auckland, D.W., Cooper, R. and Sanghera, J., 1981,
        "Photographic investigation of formative stage of
        electric breakdown in diverging fields", I.E.E. Proc.,
        128A, pp. 209-214.

3.18   O'Dwyer, J. J., 1973, ibid., pp.225-235.

3.19   Klein, N., and Solomon, P., 1976, "Current runaway
        in insulators affected by impact ionization and
        recombination", J. of Appl. Phys., 47, pp.4364-4372.

3.20   Seitz, F., 1949, "On the theory of electron
        multiplication in crystals", Phys. Rev., 73,
        pp. 1375-1391.

3.21   Klein, N., 1972, "A theory of localized electronic
        breakdown in insulating films", Advances in Physics,
        21, No.92, pp. 605-645.

3.22   Pais, J. C., 1970, "Interrelation between current,
        breakdown voltage and hardness for allyl polymers",
        I.E.E. Conf. Publ., 67, pp.101-104.

3.23   Stark, K.H. and Garton, C.G., 1955, "Electric strength
        of irradiated polythene", Nature, London, 176,
        pp.1225-1226.

3.24   Fava, R. A., 1965, "Intrinsic electric strength and
        electromechanical breakdown in polythene",
        Proc. I.E.E., 112, pp. 819-823.

3.25   Blok, J. and LeGrand, D.G., 1969, "Dielectric break-
        down of polymer films", J. Appl. Phys., 40, pp.288-
        293.

3.26   McKeown, J.J., 1965, "Intrinsic electric strengths of
        organic, polymeric materials", Proc. I.E.E., 112,
        pp. 824-828.

3.27   Lawson, W.G., 1966, "Effects of temperature and
        techniques of measurement on the intrinsic electric
        strength of polythene", Proc. I.E.E., 113, pp.197-202.

3.28   Sabuni, H. and Nelson, J.K., 1977, "The electric
        strength of copolymers", J. Materials Sci. 12,
        pp.2435 - 2440.

3.29   Hirai, N. and Eyring, H., 1959, "Bulk viscosity of
        polymeric systems", J. Polym. Sci., 37, pp. 51-70.

3.30.   Artbauer, J., 1966, "Some factors preventing the
        attainment of intrinsic electric strength in
        polymeric insulation", Acta. Tech., C.S.A.V., Prague,
        3, pp. 429-439.

3.31   McCrum, N. G., Read, B.E. and Williams, G., 1967,
        "Anelastic and dielectric effects in polymeric
        solids", Wiley, New York,

3.32   Mason, J.H., 1955, "Breakdown of solid dielectrics in
        divergent fields", I.E.E., Monograph, No.127,

3.33   Bahder, G., Dakin, T. and Lawson, G.H., 1974,
        "Analysis of treeing type breakdown", C.I.G.R.E.
        paper 15-05

3.34   Eichorn, R.M., 1976, "Treeing in solid extruded
        electrical insulation", I.E.E.E., Trans., EI-12, No.1
        pp. 2-18.

3.35   Auckland, D.W., Cooper, R. and Sanghera, J., 1981,
        ibid.

3.36   Auckland, D. W., Gravill, N., and Thomas, H. W.,
        1976, "The intensification of underexposed
        photographs of pre-breakdown luminescence in
        dielectrics", J. Phys. E., 10, pp. 132-133.

3.37   Auckland, D.W., Borishade, A.B. and Cooper, R., 1977,
        "The development of electrical discharges in
        simulated tree channels".,
        I.E.E.E. Trans., EI-13, p.113, I.E.E.E. Trans., EI-12,
        pp.349-354.

3.38   Budenstein, P., 1980, "On the mechanism of dielectric
        breakdown of solids", I.E.E.E. Trans., EI-15, pp.225-
        240.

Chapter 4

Breakdown in composites

N. Parkman

4.1 INTRODUCTION

It is difficult to conceive of a <u>complete</u> electrical
insulation system which does not consist of more than one
insulating phase. If the insulation system as a whole is
analysed, it will invariably be found that more than one
insulating material is involved. These different phases may
be in parallel with one another - such as air in parallel
with solid insulation - or, less commonly unless purposely
designed, in series with one another.

Apart from the 'accidental' composite nature of an
insulation system which arises, as described above, from the
mechanical requirements involved in separating conductors at
different potentials, even parts of a system that are
normally composed of a single material or phase are in fact
composite in nature. This arises because, in practice, single materials will inevitably have small, even microscopic,
volumes of another phase of material present in their bulk
form. A solid will contain gas pockets perhaps of exceedingly small dimensions; a liquid or gas will contain adventitious matter notwithstanding the high level of care taken
in the manufacture of such high quality insulating materials.
On the whole this chapter is concerned with composite dielectric systems which are designed specifically to enhance the
performance of that part of the dielectric or insulating
system which has to bear high electrical stresses. Specially formulated composite dielectrics of one or another
kind are very widely used in electrical apparatus because of
the practical difficulties of combining all the requisite
characteristics, electrical and mechanical, in any one
material. Among these combinations, that of thin sheets of
a flexible solid, cellulosic paper or plastic, together with
a mobile impregnating liquid, naturally occurring or synthetic, is one of the most important. Provided the application
is such that the liquid component can be contained and also
maintained, preferably under hydraulic pressure, free from
all contamination by ions, particles or solution of gases,
it provides an insulation which is not only relatively cheap
and easily applied, but also, under optimum conditions, has
the best practically available combination of high operating
stress, low dielectric loss, and slow rate of deterioration
in service. These attributes account for its present

widespread use in a great range of both high and low voltage apparatus: cables, capacitors, transformers, oil-filled switchgear and their associated bushings. In the past decade or so the increasing use of impregnating and potting resins has extended this solid/liquid technology to other areas of electrical design, but the thin insulating film impregnated with a mobile liquid still remains supreme in terms of the ability to support the maximum electrical stress over long periods of service.

The desirable properties of this kind of insulation, however, are not to be obtained, or not to their full extent, without potential penalty. Owing to the presence of a liquid component, bubbles or cavities will readily form if gas is available from outside the system, or if gas becomes available locally as a result of a number of processes, both chemical and physical, which can occur in a contaminated system. All these processes, which depend upon the movement of charges, either ions or charged particles, through the insulation system, are accelerated by the high mobility of the carriers in the liquid component. The result is that an impregnated system prepared without sufficient care against contamination may be subject to several mechanisms of very slow deterioration in service, only some of them detectable by any short-time test which can be performed on the complete apparatus.

4.2 FACTORS LEADING TO REDUCTION OF ELECTRIC STRENGTHS OF MATERIALS

In order to understand the need for composite insulation, it is necessary to elucidate the factors which control the considerable spread in the measured electric strength of single materials. Previous chapters have discussed the electric strength of 'pure' materials and the techniques necessary to ensure that the materials tested are free as possible from defects and that the test methods do not introduce extraneous factors which could cause a reduction in breakdown strength. In gases it has been shown (chapter 1) and will be discussed again in chapters 6 and 10, that the presence of particles and other features causing high stress regions can bring about large reductions in a measured breakdown voltage. Indeed the electric strength of gases can be reduced by orders of magnitude by introducing interfaces, water and particulate pollution such as may arise in insulator flashover. In liquids it has been shown (chapter 2) that particles and contaminants lead to a reduction and extreme variability in measured electric strength, with electrode effects playing an important role. Theories of conduction in 'pure' liquids exist and the processes are reasonably well understood, but when solid material is introduced into a liquid, contamination increases and the paths of ions are limited by solid dielectric barriers, producing complex charge and field distributions, with both d.c. and a.c. fields. In solids it has been shown (chapter 3) that stress concentrations associated with defects can reduce the electric strength dramatically. As a

generality it can be stated that the highest electric
strengths for all phases of materials are achieved on small
(10 mm² say) samples. As the specimen size is increased
there is a progressive decrease in observed values of electric strength (Fig.4.1). This is due to the inevitable
introduction of imperfections as the volume of the specimen
increases; these imperfections can vary from those of molecular dimensions up to gross defects arising from inclusions.
Thus if the electric strength of a thin sheet of material is
measured locally with a small electrode, it will be found
that the measured electric strength varies considerably from
place to place (Fig.4.2).J B Whitehead (4.1) writing in 1935
stated that 'the simple constituent dielectric proposed by
Maxwell does not exist'. Notwithstanding the enormous
improvements that have taken place in insulation manufacture,
this statement remains true.

The foregoing cursory treatment has been for the purpose of demonstrating that, although individual insulating
materials in their 'pure' form possess electric strengths
which are very much higher than those which are required of
them in electrical apparatus, in practice these electric
strengths cannot be utilised because of the existence of
impurities or defects in the structure of the materials.
These defects may be inherent in the material itself, may be
present due to some artefact of manufacture or may reveal
themselves during the service life of equipment due to ageing processes to be described later on.

4.3   PROPERTIES OF COMPOSITE INSULATION

In order to minimise the statistical weakness effects
referred to in the previous section the overall philosophy
in utilising composite insulation is to combine different
phases in series and sometimes in parallel, in a way which
gives superior properties and a very much higher electric
strength than would be the case for a single material of the
same thickness.

4.3.1   The Effect of Multiple Layers

The simplest form of 'composite' insulation system consists of two layers of the same material. This system is
extensively used especially in cable and capacitor construction where advantage is taken of the fact that two thin
sheets have a higher electric strength than a single sheet
of the same total thickness. The effect is particularly
marked for materials where there is a wide spread in values
of electric strength measured at different points on the
surface (as indicated earlier). In some analytical work
carried out on thin sheets of polymethylmethacrylate it was
discovered that the main factor involved in this phenomenon
was a loss of energy from the discharge or partial breakdown
channel at the interfaces. It was found in fact that a discharge having penetrated one layer could not enter the next
layer of material until the spot on the interface, centred
on the channel, had been charged to a potential which could

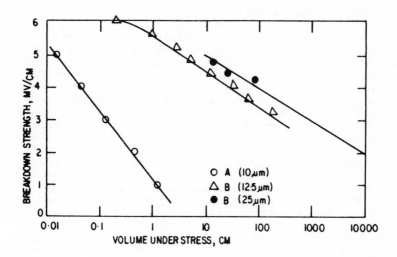

Fig.4.1   Effect of volume on electric
          strength

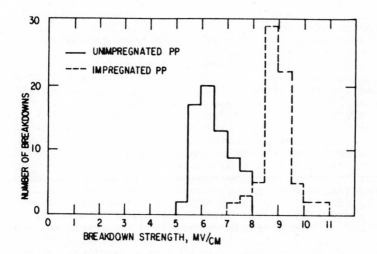

Fig.4.2   Variation in electric strength at
          different points

produce a field comparable with that of the channel at the level in question.

4.3.2 Effect of Layer Thickness

The relevance of the effect of increased electric strength resulting from a layered construction, as described above, to capacitor and cable construction is fairly obvious. It has been observed over a long period of time that breakdown channels in high voltage oil impregnated paper cables tend to wander along interfaces between layers of insulation, often for fairly long distances, in order to gain access to the next layer via a relatively weak butt-space, rather than penetrate a very much shorter path directly through the paper.

The effect of a layered construction is especially important for insulating paper since apart from the variation in electric strength across the surface due to changes in paper structure, there is also a pronounced effect due to variation in paper thickness (Fig.4.3).

Paper is relatively inhomogeneous in thickness across its width; the actual thickness at given points along a width may vary considerably. This variation has certain advantages and disadvantages. On the one hand, differences of thickness impart a rough surface to the paper which assists in its impregnation when tightly wound. This is particularly the case for the high density papers which lack any significant porosity in the body of the paper. The surface roughness then serves to assist impregnation by capillary action at the interface between adjoining layers. The disadvantage of the existence of thin spots in the paper is that the voltage to cause breakdown at these points will be considerably less than for the thicker parts of the paper. The influence of paper structure on the performance of impregnated paper power capacitors has been extensively studied by Church and Krasucki (4.2).

It is known that the discharge inception stress, E_i, of liquid-impregnated paper capacitors decreases with increasing dielectric thickness (4.3), Mason (4.4), has suggested that the effect may be a consequence of the presence of conducting inclusions in the paper. Hopkins et al (4.5) have shown that, for any given thickness of a pentachlorodiphenyl-paper dielectric, the discharge inception stress is independent of the number of sheets of paper within the limits of two to ten sheets and this suggests that the E_i-thickness dependence is unlikely to be the result of the presence of conducting inclusions in the paper. Results reported by Kutchinsky et al (4.6), show that 'low-intensity' discharges in oil-impregnated paper take place at the edges of the electrode foils. If discharges in liquid-impregnated dielectrics occur at the edges of electrodes, then it should be possible to explain the E_i-thickness dependence on the basis of variation of electric field at the edge with thickness of the dielectric.

The problem has been investigated in the following way. The permittivities of paraffin oil and silicone fluid are

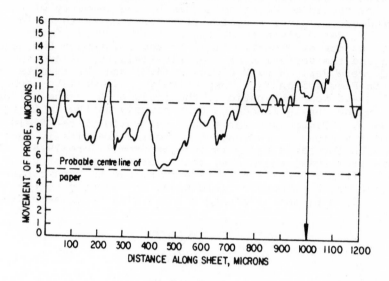

Fig.4.3   Variation in paper thickness across
the width

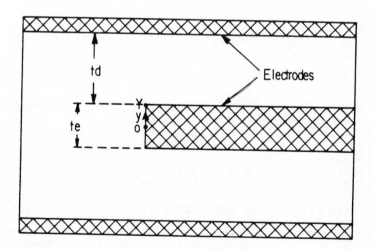

Fig.4.4   Electrode arrangement

comparable with that of polystyrene and, therefore, the distribution of electric field in impregnated polystyrene dielectrics should be determined only by the geometry of the
electrodes. Consider a system of electrodes as shown in
Fig.4.4.   For a given thickness of the electrode foils (t_e),
the dependence of electric field at the point 0 on the spacing t_d and the variation of electric field, E_y, along the
edge OY were derived by R J Clowes (4.7) and the results are
given in Figs.4.5 and 4.6 respectively.   It will be seen
from Fig.4.5 that the ratio of the field at 0 to the uniform
field in the gap t_d increases continuously as the ratio t_d/t_e
increases.   Figure 4.6 shows that the ratio of the field at
any point y along OY to that at 0 remains practically un-
changed as the ratio t_d/t_e is changed from 1.5 to 19.5.   It
follows from this, therefore, that over the range of values
of t_d/t_e from 1.5 to 19.5 it is permissible to assume that
the ratio of the electric field E_y at any point y to the
uniform field E_i in the gap t_d changes with t_d/t_e according
to:

$$E_i = A\ E_y \sqrt{\frac{t_e}{t_d}\ 1 + \frac{t_e}{4t_d}} \qquad \dots\dots\dots\dots (4.1)$$

where A is a dimensionless constant which is different for
different values of y but, in view of the results given in
Fig.4.6, can be taken as independent of t_d/t_e.   It will be
seen from Fig.4.7 that when $t_d = 15$ µm (i.e. $t_d/t_e = 2.5$)
then $E_i = 247$ volts(peak)/µm and, therefore, from
Equation (4.1):

$$A\ E_y = 372.5 \text{ volts(peak)/µm} \qquad \dots\dots\dots\dots (4.2)$$

Substituting Equation (4.2) into Equation (4.1) gives:

$$E_i = 372.5 \sqrt{\frac{t_e}{t_d}\ 1 + \frac{t_e}{4t_d}} \text{ volts(peak)/µm} \dots (4.3)$$

The dependence of discharge inception stress E_i on the ratio
t_d/t_e given by Equation (4.3) is plotted as curve (a) in
Fig.4.7.   It will be seen that there is good agreement between Equation (4.3) and experiment and this shows that in
paraffin oil and silicone fluid impregnated polystyrene dielectrics the onset of discharges occurs when a constant
critical field is applied to the liquid at the electrode
edge.
        Results reported by Kutchinsky et al (4.6) indicate that
in oil-impregnated paper discharges also occur at the edges
of the electrode foils.   Their measurements were made on
small areas of oil-impregnated paper (capacitance varying
from 50 to 100 pF) using, on one side, a transparent electrode made by depositing a conductive coating (surface
resistivity = 100 ohms) on a glass plate.   The samples were
dried and impregnated at 130°C and at a pressure of about
0.1 mm Hg.   Kutchinsky et al report that, in their samples,
light was emitted from spots at the electrode edges at a

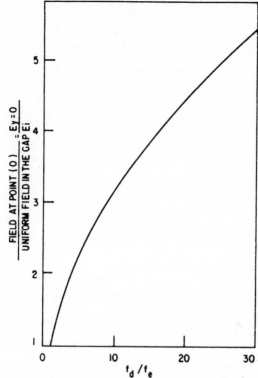

Fig.4.5   Variation of electric field at
          electrode edges

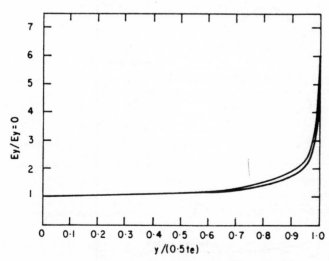

Fig.4.6   Variation of electric field at
          rectangular edge electrode

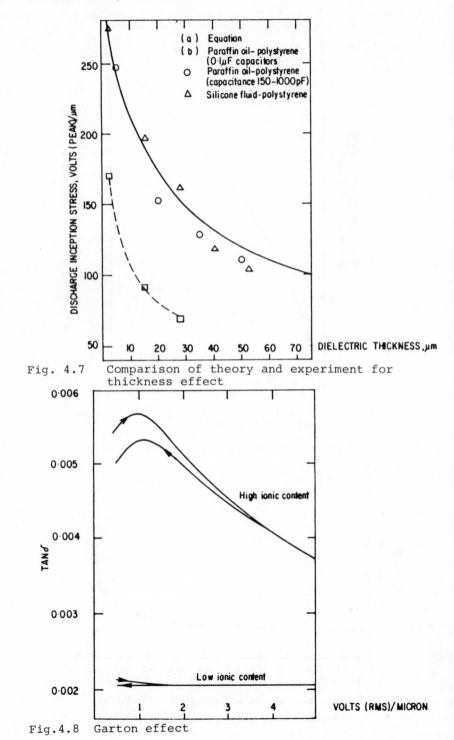

Fig. 4.7   Comparison of theory and experiment for thickness effect

Fig. 4.8   Garton effect

stress which was practically the same as the discharge
inception stress.   Their measurements showed that the variations of the discharge inception stress, E_i, and of the
light emission stress, E_l with dielectric thickness t_d are
given by -

$$E_i = E_l = 3.6 \times (t_d)^{-0.58} \quad kV/mm \quad \ldots\ldots\ldots\ldots (4.4)$$

where t_d is measured in mm $(0.03 \leqslant t_d \leqslant 1.0)$.   Substituting
$t_e = 10^{-2}$ mm in Equation (4.1) gives -

$$E_i = B \times (t_d)^{-0.5} \times \sqrt{1 + 0.0025/(t_d)} \quad \ldots\ldots (4.5)$$

where B is a constant.   It will be seen that the term
$0.0025/(t_d)$ is small compared with unity for $0.03 < t_d < 1.0$
and, therefore, Equation 4.5 becomes, approximately -

$$E_i = B \times (t_d)^{-0.5} \quad \ldots\ldots\ldots\ldots (4.6)$$

For Equation (4.6) to be valid, the permittivity of the solid
must be equal to that of the liquid.   Although this condition
is not satisfied in an oil-paper dielectric, there is satisfactory agreement between Equations (4.4) and 4.6) and this
suggests that -

(i) the difference in permittivity between oil and
    paper does not affect significantly the rate of
    change of field at the electrode edge with di-
    electric thickness,

(ii) discharges in oil impregnated paper set in when
     the electric field at the edge of the electrode
     foil reaches a certain critical value.

4.3.3 Liquid/Solid Interfaces between Layers - 'Garton Effect'

One of the most important and long established composite
insulation systems is oil-impregnated paper or, more generally
to allow for changes particularly in the past decade, insulating liquid-impregnated thin insulating films of plastic and/
or paper.
In these composite systems it is essential to maintain
low dielectric losses because of the high electric stress at
which such systems operate.
The dielectric loss of pure liquids is well understood
and is not considered here.   Where loss is critically
important in an impregnated structure, the effects are minimised by using non-polar impregnants such as mineral oil, or by
using polar liquids in the range of temperature and frequency
where the loss angle is small.   When, however, even an
initially pure liquid is used under industrial conditions to
impregnate a porous or laminated solid dielectric, impurities
capable of dissociating into ions are very likely to be

picked up either from some chance sources of impurity in the impregnating plant or, in some cases, by leaching from the solid insulation itself.   In fact, it is practically impossible to maintain such a high degree of purity that no ionic loss angle is detectable in the impregnated condition.   If the impurity happens to be one which dissociates more or less completely in solution, extremely small concentrations, down to 10^{-8} (weight fraction), in a polar liquid, will produce a measurable loss angle.   Fortunately, such extremely active substances are rare, but among them are some partly oxidized residues of organic materials, e.g. bitumen, rosin and oils, and also some size-like products used in 'finishing' textiles and pasteboards.   Since these materials are quite likely to be used somewhere in an electrical factory, great care is needed to exclude them from all areas concerned with liquid impregnation.

The most noticeable effect of moderate ionic contamination, in a composite system such as a liquid containing solid insulating barriers, is a high (typically, a maximum) value of loss angle at a stress much below the working value, followed by a sharp fall toward a much lower and nearly constant value toward higher stresses (Fig.4.8).   This behaviour seems to have been first noted by J B Whitehead (Ref.4.1) and ascribed by him probably to limited ionic motion, about 1935, in an oil-impregnated cable.   In a polar liquid, the effect can be produced more easily, and it was first investigated under controlled experimental conditions, and the mathematical theory given, by Garton (Ref.4.8) in 1941.   Since it has not acquired any brief electrical name, it is often referred to as the 'Garton-effect'.   The full theory is too long to be given here and attention is focussed instead on the use of the effect as a diagnostic tool for determining the characteristics, and hence possibly the origin, of a particular contamination.

The change of loss angle, measured at low enough stress, can be used quite generally to follow changes of ionic concentration in the liquid fraction of composite insulation as a result of some treatment to which the specimen has been subjected, even when the changes at working stress are small or confused.   For example, Plessner, Shen (Ref.4.9) used the method to study transfer of ions between liquid and plastic in capacitors of that type when much higher a.c. stresses had been applied before the measurement.   The effect was to reduce the low-stress loss angle to half or less of its initial value, most of the change taking place in the first few hours, but continuing measurably for several hundred hours.   The effect is fully recoverable on removal of stress, but at an even slower rate.   The authors attribute the effect, very plausibly, to diffusion of ions from the liquid into the plastic, during alternate half-cycles when they are held against it, in high concentration, by the action of the field.   More importantly, they point out that since ions of the other sign will similarly diffuse in during the other half-cycles, the two may recombine within the plastic, and if the dissociation constant there is smaller than in the liquid, there will in course of time be a bulk transfer of liquid into the plastic.   It is known that water-absorption by polyethylene is greatly increased in an a.c. field, and it may be that the same mechanism is involved.

4.4 MECHANISMS OF BREAKDOWN IN COMPOSITES

On the assumption that dielectric losses, arising from sources already discussed in the preceding section, are sufficiently low to obviate cumulative heating and thermal breakdown, we can now concentrate on other factors which determine the short and long time breakdown strengths.

4.4.1 Short Term Breakdown

In the high-stress case, failure may occur in seconds or fractions of seconds without any substantial damage to the insulating surfaces prior to breakdown.   For 'lossy' materials, breakdown may be due to thermal instability, except that at very short times the mechanism of breakdown is probably similar to that for low-loss material.   The effect of applying a continuously increasing voltage to samples, in the presence of discharges, has been investigated in the laboratory.   It has been shown that breakdown results from one or perhaps a very few discharges when the voltage is near to the observed breakdown value.   The investigation into this form of breakdown is in no way complete, but it appears that there exists a critical stress in the dielectric, at which discharges of a given magnitude can enter the surface of insulation and propagate rapidly to cause breakdown. Tests with single discharges on to an insulating material have shown that the critical field decreases as the discharge magnitude increases.   These tests also show that breakdown occurs more readily when the electric stress in the insulation is of a pattern which assists the bombarding particles in the discharge to penetrate the insulation than when it opposes their entry.   Breakdown also occurs more readily when the bombarding particles are electrons rather than positive ions.   These features are indicated in (Fig.4.9) which shows the variation of stress in the insulation required to cause breakdown with changing thickness of insulation when a discharge of fixed magnitude falls on the insulating surface.

There can be little doubt that one effect of a discharge impinging on a surface is to increase the stress locally. If we assume that the point of a discharge is equivalent to a conducting prolate hemispheroidal boss of length 'c' and base radius 'b', projecting into the insulation along the z-axis, then the potential along the z-axis is given by V -

$$ V = E_O z \left(1 - \frac{u - \text{arc tanh } u}{w - \text{arc tanh } w}\right) \qquad \dots\dots\dots\dots (4.7) $$

where $u^2 = \frac{(z^2 - b^2)}{z^2}$, $w^2 = \frac{(c^2 - b^2)}{c^2}$

and E_O is the uniform stress at a point remote from the boss.

The stress at a point on the z - axis is, therefore,

$$E = \frac{dV}{dz} = E_O \left(1 + \frac{u}{(u^2 - 1) + \text{arc tanh } u} \right)$$
$$\frac{}{w - \text{arc tanh } W}$$

The field intensification factor E/E for values of c/b are given in Fig.4.9, where it will be seen that for c/b $\pm$ 1, i.e. a hemisphere, the field at the tip of a discharge is intensified by a factor of three.

There can be little doubt that this local field intensification plays a vital role in breakdown under high field conditions, the actual effect depending on the field in the insulation before the discharge impinges on it.

4.4.2 Longer Term Breakdown

The principal ageing effects responsible for breakdown of composite insulation systems arise from thermal processes and the presence of partial discharges.   The question of breakdown due entirely to thermal ageing is treated elsewhere (Chapter 12).

4.4.2.1 Breakdown by discharges in cavities.   One of the most important factors in determining the electric strength of any practical composite dielectric is the fact that it is difficult to manufacture apparatus which is free from discharges in gas-filled cavities within the dielectric or adjacent to the exposed interface between conductor and dielectric.   The discharges involve the transfer of charge between two points in sufficient quantity to discharge the local capacitance.   At a given voltage, the impact of this charge on the dielectric surface produces a deterioration of the insulating properties in diverse ways depending on geometrical factors and the nature of the dielectric.   Since breakdown by discharges is so important industrially, it is worth while examining the factors which control the discharge inception voltage V_i, and the discharge magnitude. It is interesting to note the effect of permittivity ε and depth 'g' of a discharging cavity on the changes in V_i as the cavity increases in depth.

It is easy to show that V_i is given approximately by:

$$V_i = \frac{E (t + \varepsilon g)}{\varepsilon} \qquad \dots\dots\dots\dots (4.8)$$

where E is the breakdown stress of the air gap and 't' is the total thickness of dielectric in series with the gap.   For a given arrangement (g + t) will be constant + C.   Then equation 4.8 may be written,

$$V_i = E \frac{[C + (\varepsilon-1)g]}{\varepsilon} = \frac{C}{\frac{(\varepsilon-1) + g}{\varepsilon}} \qquad \dots\dots\dots\dots (4.9)$$

Differentiating,

$$\frac{dV_i}{dg} = \frac{\varepsilon - 1}{\varepsilon} \left[E + \left(\frac{C}{(\varepsilon-1) + g} \right) \frac{dE}{dg} \right] \quad \ldots \ldots \ldots \ldots \quad (4.10)$$

Now E is always positive and dE/dg always negative or zero. For small 'g', dE/dg increases more rapidly than E with decreasing 'g' and Equation (4.10) is negative, but dV_i/dg can still be negative for larger 'g' provided $C/(\varepsilon-1)$ is large enough.

Thus for very small cavities, V_i decreases as the cavity depth increases. The extent of the decrease depends on the value $C/(\varepsilon - 1)$ as shown in Fig.4.10. The full curves are for C = 2.3 mm, and the dotted curves for C = 4.3 mm. Both families of curves are for ε in the range 2 to 8.

It will be seen that for $\varepsilon = 2$, C = 2.3 and initial g = 0.3 mm, V_i is more or less the same initially as when the whole of the dielectric surrounding the cavity has been eroded. For dielectrics with $\varepsilon = 6$, V_i is doubled as the dielectric is eroded. Superficially, it is therefore difficult to see how breakdown could occur by simple erosion of those materials at voltages below $2V_i$. In fact, it is found that at voltages less than $2V_i$, life is indeed very long. The total capacitance of a cavity is not discharged as a single unit but as a result of many discharges, each involving only a small area determined by the conductivity of the cavity surface in the region of the discharge. Neither is the onset of all the discharges simultaneous, but succes- sive as a rising voltage provides the necessary stress to attain local breakdown conditions.

4.4.2.2 Surface effects. It has been found that for cavities 0.3 mm deep enclosed in polythene, the discharge magnitude does not continue to increase when the discharging faces are increased above 0.3 cm diameter. Tests carried out since suggest that if the discharges occur between spaced surfaces this is true only for uncontaminated samples. When the surface conductivity increases, the discharge magni- tude increases also, and with it the damage caused to the dielectric. Kanonykin (4.10) made similar observations for discharges between glass plates. He also observed directly that by increasing the surface conductivity with semi- conducting paints, the discharge changed from a diffuse glow to concentrated sparks.

It is useful to examine quantitatively the parameters affecting the discharge magnitude between dielectric surfaces of a given form. The largest area which can be discharged instantaneously in a single discharge is dependent on the associated surface resistance and capacitance given a time constant of the same order as the discharge duration. Consider a discharge radius 'r' which takes charge from a surface of radius R. Assuming the resistivity p_s of the discharging surface to be uniform, the total resistance R_s between two concentric rings of radii R and 'r' is -

$$R_s = \frac{p_s}{s} \log_e \frac{R}{r} \quad \ldots \ldots \ldots \ldots \quad (4.11)$$

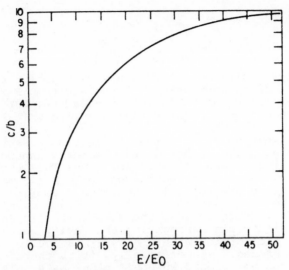

Fig.4.9   Field intensification due to geometrical factors

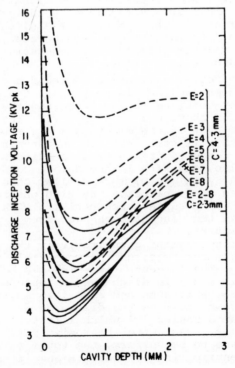

Fig. 4.10   Variation of discharge inception with cavity depth

The local capacitance associated with two parallel plates of radii R separated by a thickness 't' is

$$C_s = \frac{1.1\pi\ R^2\ 10^{-12}}{4\pi\ t}\ \ F \qquad \ldots\ldots\ldots\ldots (4.12)$$

The time constant associated with the discharge is given by the product of the expressions 4.11 and 4.12. By equating this to γ, the discharge duration, the radius (R) of the area discharged is given by

$$1.1\ R^2\ 10^{-12}\ p_s\ \ln\ (\frac{R}{\frac{r}{8\pi t}}) = \gamma \qquad \ldots\ldots\ldots\ldots (4.13)$$

Measurements of the spectrum of discharges between polythene surfaces indicate times (γ) of the order 10^{-6} to 10^{-7}s. Taking equation 4.13 we now write

$$R^2\ \ln\ \frac{\frac{R}{r}}{r^2} = \frac{8t\ 10^5}{1.1\ p_s\ r^2} \qquad \ldots\ldots\ldots\ldots (4.14)$$

and taking $\gamma = 10^{-7}$s

i.e. $$x^2\ \ln\ x = \frac{8\pi t\ 10^5}{1.1\ p_s\ r^2} \qquad \ldots\ldots\ldots\ldots (4.15)$$

where $x = \frac{R}{r}$

Figure 4.11 is a plot of $x^2 \ln x$ for various values of 'x'. Taking 'r' the radius of the discharge channel as 10^{-2} cm the plotted values of 'x' represent values of R from 10^{-2} to 1 cm. Assuming the value of t = 0.03 cm used in experiments, the values of p_s corresponding to a given R are given along the abscissa.

For values of $p_s > 10^7$, R is given by 4.15 as 10^{-1} cm, whereas in practice it is known from electro-photography and other measurements that transverse discharges occur on the faces of the dielectric causing larger areas to be discharged instantaneously. With increase in surface conductivity the transverse stress will decrease and the radius of the area discharged will be given by Equation 4.15.

Changes in surface conductivity are brought about by the discharges themselves so that the changes in discharge magnitude occur spontaneously during the life of a dielectric. If the tests are carried out in air, deliquescent nitrates are produced on the dielectric surface, e.g. calcium nitrate formed on materials containing cellulose fillers, since calcium salts form the major part of the inorganic residues contained in cellulosic materials.

It is well known that when discharges occur in cavities enclosed in dielectric, or between spaced surfaces, uniform erosion of the dielectric surface occurs at first. After a time, dependent on the applied voltage and frequency, a 'weak spot' is formed or the discharges 'concentrate' on a few points or a single point, causing rapid erosion and breakdown. The onset of this 'weak spot' phenomenon is also known to be associated with the occurrence of larger discharges, but cause and effect have always been confused. The observed change of discharge characteristics with life under the impact of small discharges could now be explained as follows. For clean surfaces at the discharge inception voltage the discharge pattern is characteristic of the dielectric and its form. It will consist normally of a number of comparatively small discharges from sites on the surface where the necessary discharge conditions exist. After a little while, erosion at these sites causes the discharges to decrease in number and magnitude and total extinction may occur. As life continues, the position becomes complicated for enclosed cavities by the co-existence of opposing tendencies resulting from discharge-induced surface conductivity.

On the other hand, increase in surface conductivity permits larger areas to discharge instantaneously. At the same time increased surface conductivity reduces the stress across the cavity at a given voltage. If the voltage level is sufficient to maintain discharges, as surface conductivity increases the number of discharges per cycle will decrease and the discharge magnitude increase. At a critical surface conductivity the glow characteristic changes to an arc condition and the whole of the surface may be discharged instantaneously. The damage done to the dielectric increases sharply, in some cases rapidly precipitating breakdown. Where a cavity is totally surrounded by a material lacking significant quantities of either of the elements of water, e.g. polythene, there may be no appreciable increase in surface conductivity until the onset of surface carbonisation.

REFERENCES

4.1   Whitehead, J.B., 1935,
      'Impregnated Paper Insulation', John Wiley and Sons, Inc.

4.2   Church, H.F. and Krasucki, Z, 1966,
      CIGRE Paper No. 112.

4.3   Kutchinsky, G.S., Renne, V.T. and Fainitsky, V.M., 1954,
      Electrichestvo, 6, 70.

4.4   Mason, J.H., 1958,
      ERA Report No. L/T 373.

4.5   Hopkins, R.T., Walters, T.R. and Scoville, M.E., 1951,
      AIEE Trans 70, 1643.

4.6   Kutchinsky, G.S. et al, 1966,
      CIGRE Paper No. 137.

4.7   Clowes, R.J., 1967,
      Addendum to ERA Report 5197.

4.8   Garton, C.G., 1941,
      J. IEE, 88, Part II, 103-120.

4.9   Plessner, K.W., McNicholl, R. and Shen, T.E., 1974,
      Proc. IEE, 121, 1599-1602.

4.10 Kanonykin, B., 1939,
      Zh. Tech. Phys. USSR, 9, 876.

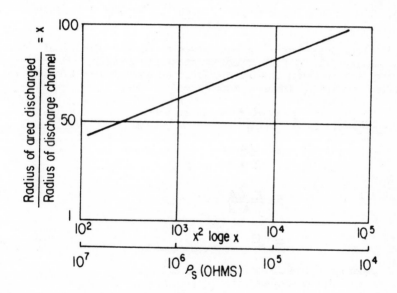

Fig.4.11   Radius of area discharged for
           given surface resistivity

Chapter 5

The electrical properties of dielectrics

W. Reddish

5.1 INTRODUCTION

The classification of substances as either conducting or insulating is familiar. Conductors follow Ohm's Law,

$$e.g. \quad I = \frac{V}{R} \quad or \quad Current \ (Amps) = \frac{Potential \ (Volts)}{Resistance \ (ohms)} \quad (5.1)$$

From this there are a series of easy steps to the calculation of specific resistivity ρ for a specimen of simple geometry and expressing this in terms of electric field strength E (volts/metre) and current density j (amperes/metre^2) :-

$$\rho = \frac{R.A}{d} \quad (ohm. \ metre) \quad (5.2)$$

$$= \frac{V.A}{I.d}$$

$$= \frac{E.d.A}{j.A.d.}$$

$$= \frac{E}{j} \quad (ohm. \ metre) \quad (5.3)$$

where A = cross sectional area (m^2)
d = thickness (m)

N.B. we have assumed (i) R and ρ are constants ie not varying with time.

(ii) that the material is both isotropic and linear.

Resistivity measurements are made readily with conductors and semi-conductors. We now ask the question - do insulators follow the same law but with large R values? In attempting to answer this question we find that it is necessary to measure much smaller currents than are found for conductors or semi-conductors. In the past sensitive galvanometers, quadrant electrometers, quartz fibre electrometers or valve electrometers have been used for such studies (5.1). Nowadays a wide range of small current measurements can be

transient he cause

$C_0 + C(t)$

made with negative feedback electrometer circuits such as in Fig 5.1.

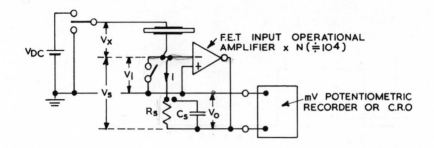

Fig.5.1   Schematic Electrometer Amplifier

$$V_o = -NV_i \qquad (5.4)$$

$$V_x = V_{DC} - V_i = V_{DC} + \frac{V_o}{N} \doteq V_{DC} \qquad (5.5)$$

$$V_s = V_i - V_o = -\left(V_o + \frac{V_o}{N}\right) \qquad (5.6)$$

$$i = \frac{V_s}{R_s} = -\frac{1}{R_s}\left(V_o + \frac{V_o}{N}\right) = -\frac{V_o}{R_s}\left(1 + \frac{1}{N}\right) \doteq -\frac{V_o}{R_s} \qquad (5.7)$$

As the electrometer amplifier has a very large input resistance, R_s can be as large as 10^{13} ohms.

E.g.   If $V_o = 100\,mV$, $R_s = 10^{12}\,ohms$, $V_{DC} = 100\,v$

then $i = 10^{-13}\,A$ and $R = \frac{V_{DC}}{i} = 10^{15}\,\Omega$

5.2 D.C. RESPONSE

5.2.1 Polarisation

If we use such equipment, apply volts to a typical good insulator and measure current we do not get simple Ohm's Law behaviour. We find that, as illustrated in Fig 5.2, the specimen 'charges up' or is polarised when volts are applied, drawing relatively large transient currents in the process and is 'discharged' or depolarised when the voltage is removed with transient current flow in the opposite direction. We have designated the charging current as $i(t)$ i.e. i as a function of time. These polarisation and depolarisation processes can be very fast for some materials. e.g. microseconds; for others they often take a long time (e.g. hours or days). Any residual steady state current is much smaller than the transient current and is often not resolved even with the most sensitive electrometers available. This is the norm rather than the exception for organic polymer dielectrics at room temperature.

In practice it is necessary to short circuit the input of the amplifier (Fig 5.1) at the beginning of the charging and discharging processes in order to avoid damage to the input circuit. Also R_S values, over a logarithmic range (e.g. 10^6 to $10^{13}\,\Omega$), can be selected in order to accommodate the large range of current values involved in this type of measurement.

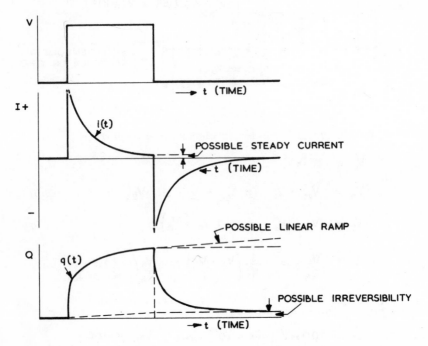

Figs.5.2   Charge/Discharge or d.c. Step response behaviour

The $q(t)$ curve in (Fig. 5.2) can be obtained by replacing R_S in (Fig. 5.1) by the reference air capacitor C_s. (e.g. $C_s = 100\ \text{pF}$) in which case we have, in place of Equation 5.(7):-

$$q(t) = C_s V_s = -C_s\left(V_o + \frac{V_o}{N}\right) = -C_s V_o\left(1 + \frac{1}{N}\right) \doteq -C_s V_o(t) \qquad (5.8)$$

$$\text{and} \quad i(t) = C_s \cdot \frac{dV_s}{dt} = -C_s \frac{dV_o}{dt} \qquad (5.9)$$

Thus a steady state current $i(t)$ = constant, if present, produces a linear ramp voltage output and a corresponding irreversibility in the discharge curve. When using a reference capacitor it is not necessary to short circuit the amplifier input during the initial transient and all the charge is recorded.

It is sometimes advantageous to use both R_S and C_s together, particularly at large polarisation times as a means of increasing signal/noise ratio. In which case we have:-

$$i(t) = -\left(\frac{V_o(t)}{R_s} + C_s \cdot \frac{dV_o}{dt}\right) \qquad (5.10)$$

5.2.2   Permittivity

In general $Q = CV$ (5.11)

For a (parallel plate) capacitor with vacuum between the plates.

$$C = C_g = \frac{A \cdot 10^{-10}}{d \cdot 3 \cdot 6 \, \pi} \; \text{Farads}$$ (5.12)

Cg is a constant attained instantly. (A and d in metric units)

With a dielectric filling the space between the plates we have:-

$$C = \varepsilon(t) \, Cg$$ (5.13)

The parameter $\varepsilon(t)$ contains both the magnitude of the increase in capacitance and the time dependence due to the presence of the dielectric.   Its value can be obtained directly from $q(t)$ measurements using (Fig. 5.1) with reference capacitor and Eqns (5.8), (5.11), (5.12), (5.13).

$$\varepsilon(t) = \frac{C}{C_g} = \frac{1}{C_g} \cdot \frac{q(t)}{V_{DC}} = -\frac{C_s}{C_g} \cdot \frac{V_0(t)}{V_{DC}}$$ (5.14)

In (Fig. 5.4) values measured in this way are quoted[5.3] for pure polyvinyl chloride measured over a range of temperatures.   Values ranging from 3 to 20 were recorded.   Only at temperatures over 95°C was a steady state value or static dielectric constant reached after 1000 seconds polarisation.

(*   Negative sign here ensures that $\varepsilon(t)$ is a positive quantity; $V_0(t)$ is - ve).

For a conductor, the current response is constant instantly on application of voltage and we have resistivity $\rho = \frac{E}{j}$ (Eqn 5.3) = constant.

For a dielectric, because current response is time dependent, E/j is not a constant. It is however useful to quote measured E/j values as a function of polarisation time, and to reserve the term resistivity - the volume resistivity encountered in specification testing - to cases where $j(t)$ has become constant. (Fig. 5.3) is an example where the results of d.c. step-response measurement on a plasticised PVC compound over a range of temperatures were quoted in this way (5.2).

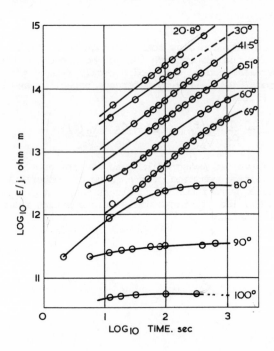

Fig.5.3   D.C. step function response data for 4.5% DOP P.V.C. compound.

It is noteable that no constant values of E/j were observed for this material below 80°C although polarising times up to 1000 seconds were employed. Typically currents reduced by a factor 50 between 10 secs and 1000 secs after voltage was applied.

We have stated as experimental fact that charging times can vary enormously for different dielectrics. The more ideal the dielectric, the more closely is the charge Q (Coulombs) reached instantaneously with the application of voltage V and the smaller is the residual steady state conduction current.

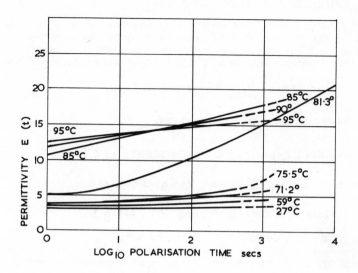

Fig.5.4   Time dependent permittivity of
polyvinylchloride at various temperatures.

From 5.11, 5.12 and 5.13 we have:-

$$Q = \frac{\varepsilon(t) \cdot A \cdot 10^{-10} \cdot V}{d \cdot 3 \cdot 6\pi} \qquad (5.15)$$

$$= \frac{\varepsilon(t) \cdot A \cdot 10^{-10} \cdot d \cdot E}{d \cdot 3 \cdot 6\pi}$$

Hence $\dfrac{Q}{A} = \varepsilon(t) \cdot \dfrac{E \cdot 10^{-10}}{3 \cdot 6\pi}$ *Coulombs/m²*

The quantity $\varepsilon(t)$ E=D is called the dielectric displacement

$$D = \varepsilon E = \frac{3 \cdot 6\pi}{10^{-10}} \cdot \frac{Q}{A} \qquad (5.16)$$

5.3 A.C. RESPONSE

5.3.1 Complex Permittivity and Loss Angle

For a large majority of engineering purposes we need to know the response of dielectrics to alternating voltages. To investigate this we measure in various ways the magnitude and phase of the current through our capacitor specimens subject to an alternating voltage.

$$(\text{vector})\quad \underline{V}_{ac} = V_p \sin \omega t \qquad (5.17)$$

$$\text{we have}\quad i = \frac{dq}{dt} = -\frac{d(cv)}{dt}$$

$$\text{for an ideal capacitor } C = \text{constant eg } C = C_g$$

$$\text{Hence}\quad i = C_g \cdot \frac{dV}{dt} = V_p . C_g . \omega \cos \omega t \qquad (5.18)$$

$$\text{where (vector)}\ \underline{I} = \underline{I}_Q = j\omega C_g \underline{V}$$

which is in quadrature with $\underline{V}$, leading by 90°.

For a general non-ideal capacitor there is an in phase component of current Ip in addition to this quadrature component I_Q and thus a complex representation of the specific dielectric parameter is required.

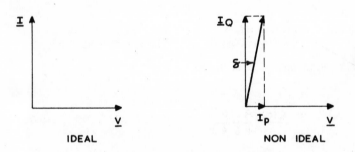

Fig.5.5   Capacitor AC Response Vector Diagram

$$
\begin{aligned}
\underline{I} &= \underline{I}_Q + \underline{I}_P \\
&= I_Q (1 - j \tan \delta) \\
&= j\omega C_g \varepsilon' (1 - j \tan \delta) V \\
&= j\omega C_g V (\varepsilon' - j \varepsilon'') \\
&= j\omega C_g V (\varepsilon^*)
\end{aligned}
\right\} \qquad (5.19)
$$

ε^* is termed the complex permittivity with ε' and ε'' its real and imaginary components. ε^*, ε' and ε'' are in general functions of ω. δ, the angle of departure from ideality, is termed the loss angle and $\tan \delta$ is known as the loss tangent which is also a function of ω.

The rate at which energy, which is stored as potential energy of polarisation of the dielectric, is converted into heat as given by:

$$\text{Power loss} = W = V.I_p$$

$$= V^2 C_g \, \omega \, \varepsilon''(\omega)$$

$$= \frac{E^2. d^2. A}{10^{10}. 3 \cdot 6 \, \pi. d} . 2\pi f. \varepsilon''(\omega) \qquad (5.20)$$

$$\frac{W}{\text{Volume}} = \frac{E^2}{1 \cdot 8. 10^{10}} . \varepsilon''(\omega) f \quad watts/m^2$$

where V, E, Ip are rms values.

These are the equations of dielectric heating.

The d.c. step response parameters $i(t)$, $E_{j(t)}$ or $\varepsilon(t)$ and the a.c. response parameters $\varepsilon^*(\omega) = \varepsilon' - j\varepsilon''$ are related via the Fourier Integral transform:-

$$\varepsilon^*(\omega) = \varepsilon'(\omega) - j\varepsilon''(\omega) = \varepsilon'(\omega)\left(1 - j\tan\delta(\omega)\right)$$

$$= \frac{1}{C_g V} . \int_0^{\infty} i(t) e^{j\omega t}. dt. \qquad (5.21)$$

It is easy to see that if loss angle and hence ε'' are small over a wide frequency range, this corresponds to $\varepsilon'(\omega)$ being approximately constant independent of frequency. In turn this corresponds to having a very small polarisation time and remaining independent of time subsequently. In this limiting case, from Equations 5.13 and 5.18,

$$i = \frac{d.(cv)}{dt} \doteqdot \varepsilon(t). C_g. \frac{dV}{dt}$$

$$\doteqdot j\omega \, \varepsilon(t) C_g. \underline{V}$$

$$\doteqdot j\omega \varepsilon' C_g. V \qquad (5.22)$$

$$\text{ie} \quad \varepsilon(\omega) \doteqdot \varepsilon(t) \doteqdot constant$$

Now such relatively ideal dielectrics exist and are used in large quantities in electrical engineering. These materials also obey Maxwell's Law

$$\varepsilon = n^2 \tag{5.23}$$

where ε is a low frequency (1 kHz) permittivity and n is the visible region (5 x 10^14 Hz) refractive index.

For example, loss angle values for the low loss polyolefins - polyethylene and polypropylene and the fluorocarbon polymer polytetrafluoroethylene are shown in (Fig. 5.6).

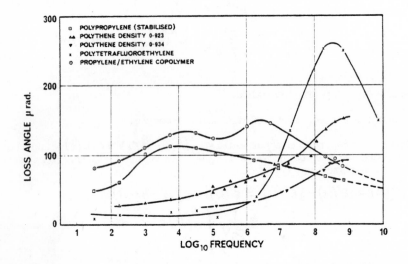

Fig.5.6   Loss angle at room temperature for some low-loss polymers.

These results were obtained using a Transformer Ratio Arms Bridge in the audio frequency region up to 10^5 Hz. and resonant circuit and cavity equipment in the radio and microwave regions. (5.4, 5.5, 5.6, 5.7).

Now the change of the real part of the complex permittivity $\Delta \varepsilon'$ is related to the area under the loss curve by the (Kramers-Krönig) relation:-

$$\Delta \varepsilon' = \frac{2}{\pi} \int_{-\infty}^{+\infty} \varepsilon'' \, d\left(\log_e \omega\right) \tag{5.24}$$

so for the materials in (Fig. 5.6) the total increase of ε' over the 10 decades of log frequency is approximately 0.002.

(The Kramers-Krönig relation is one consequence of the fact that ε' and ε'' are not independent of each other as they are both derivable from a single i(t) curve via the Fourier Integral Eqn 5.2.1.

5.3.2 Polarisation Mechanisms

The visible region refractive index n, is well accounted for by an additivity scheme of bond polarisabilities,[5.8], i.e. an electronic property, and the Clausius Mossotti formula:-

$$\frac{n^2-1}{n^2+2} = \frac{p \cdot \sigma}{M} \tag{5.25}$$

where p is the sum of the bond polarisabilities for a molecule (or repeat unit)

M is the molecular weight (or repeat unit weight)

σ is the density

The agreement between measured low frequency ε' values for polyethylene and values of n^2 calculated from Equa. 5.25 using Denbigh's polarisability values was within 1%. ε' values can be measured, using liquid immersion methods with 1 centistoke siloxane fluid as reference liquid, to an accuracy of 0.001.[5.6] ε' values obtained for both polyethylene $\varepsilon' = 2.280$ at $\sigma = 0.920$ and polypropylene, $\varepsilon' = 2.24$ at $\sigma = 0.90$, are closely related to sample density via Eqns 5.23 and 5.25 with $p/M = 0.325$. This density dependence also accounts for variation of ε' for these materials with temperature and pressure. These relations are therefore very significant in the design of ocean telephone cables where the precise value of ε' at sea bottom temperature (3°C) and pressure must be predictable.

Thus the electronic polarisation which accounts for the visible region refracture index is also the dominant contribution to the low frequency (static) permittivity for these low loss materials. It is seen as an equilibrium response which is achieved in a very short time (10^{-15} sec). It can be considered (classically) as the average displacement by the field of the centroid of all the bound electrons with respect to that of the bound positive (nuclear) charges.

We now consider other mechanisms involving the average (reversible) displacement of bound charges when field is charged. The most obvious is the relative displacement of charged atoms, ions, particularly in ionic crystalline substances (e.g. an Na^+Cl^- crystal). This is termed atomic polarisation. In this case, as in the electronic case, the charges in the absence of the field are held in equilibrium positions by the local molecules or crystal interaction potential field and are displaced from these equilibrium positions by the field. Now there exist natural frequencies of oscillation about the equilibrium positions and these oscillations are excited when the frequency of the electric field or electromagnetic wave corresponds to the natural frequency. This type of interaction produces the characteristic resonant dispersion/absorption response and is the basis of ultra violet, visible, infra-red and microwave spectroscopy. These resonant mechanisms have been treated theoretically as quantised transitions between electron or phonon energy states.[5.9, 5.10]

A most important polarisation mechanism is that involving the reorientation of permanent molecular dipoles in the electric field. This is the dominant mechanism found for organic substances including many polymers. The examples in (Figs. 5.3) and (5.4) belong to this class. The simplest cases are described by the Debye equations, discussed later, so the phenomena are often called Debye absorptions.

The last type of mechanism to be considered, the so called interfacial polarisation, is observed for heterogeneous substances. Mobile conduction charges are held up at some boundary within the overall dielectric e.g. a layer boundary or a phase boundary so that polarisation is macroscopic. An electrolytic capacitor where charge is held up at the electrode surface double layer could be regarded as an extreme member of this class. Such materials are usually discussed in terms of lumped resistor and capacitor circuit models.

There are thus three main categories of polarisation mechanism which are all slower than electronic polarisation. All of these polarisations contribute in varying degrees to the low frequency (static) permittivity of those dielectrics which are less ideal than those which follow Maxwell's relation (Eqn. 5.23), and each exhibits characteristic dispersion/absorption regions in the total spectrum. These mechanisms may be resonant mechanisms (electronic and atomic polarisations) or relaxation mechanisms (dipole orientation and interfacial polarisation). The four contributions to low frequency permittivity, the characteristic dispersion regions and their associated absorption or dielectric loss peaks are depicted schematically in (Fig. 5.7).

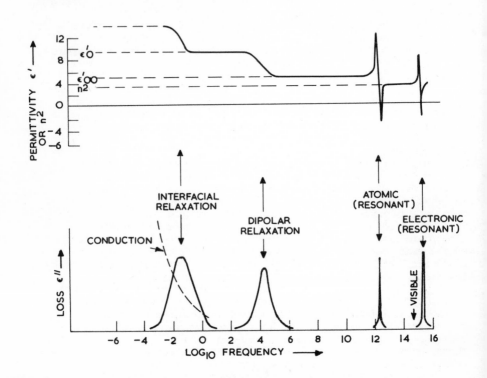

Fig.5.7   Schematic spectrum of dielectric polarisation mechanisms.

An example of a relatively simple atomic polarisation contribution is found in recent data for pure fuzed quartz. Parker et al[5.11] measured directly, in the 20 to 120 cm^{-1} region (6 x 10^{11} to 3.6 x 10^{12} Hz), using the technique of dispersive Fourier Transform Interferometry with a Michelson Interferometer, the real and imaginary parts of the complex refractive index. The data showed the character of the low frequency tail of a resonant polarisation mechanism. A curve fitting extrapolation to zero frequency yielded $n^2 = 3.8068$. A precision audio frequency a.c. bridge measurement on the same sample yielded $\varepsilon' = 3.8075$. The value of the visible region $n^2 = 2.1025$ so the atomic polarisation contribution is 1.7043.

Variants of the same Fourier Transform Interferometric methods are being used to delineate in detail the shape of the ε' and ε'' curves through the "Restrahlen" region where the alternating electric field of the electromagnetic wave is in resonance with the crystalline lattice vibrations. A good example of this work for K.Cl.crystals was published in the most recent IEE Dielectrics Conference[5.12]. The effect of cooling to 4°K was to narrow and sharpen the dispersion/absorption.

5.3.3 The Effect of Frequency

If we consider some of the implications of (Fig. 5.7), we realise what powerful methods are available to measure dielectric properties. In principle, the whole of this spectrum is available to experimental observation. Overall frequency coverage would at present require quite a large number of separate pieces of apparatus, but significant attempts are being made using new electronic devices and materials to cover larger and larger segments with single equipments. The response of a specimen, in the form of a capacitor, to alternating voltage is linked via Maxwell's equations to the response of the dielectric medium to electromagnetic waves so that electrical and optical responses constitute a continuous progression. The extension of at least 6 decades in the low frequency direction via the Fourier Transform of d.c. voltage step response has proved valuable as will be discussed further.

We need only to recognise that all this frequency range of measurements can be applied to specimens held at temperatures over a wide range. Pressure can be varied from hard vacuum to the Bridgeman range (e.g. 100 000 Atmospheres). Thus matter in a wide selection of its possible states - gaseous, liquid, glassy or crystalline solid - is accessible to dielectric study. In particular the change of dielectric properties occurring at changes of state can be measured; such measurements have contributed to understanding of atomic and molecular behaviour at such transitions. The literature on this subject is vast and has accrued at an increasing pace since the turn of the century, being linked with many important developments in chemistry, physics and engineering.

The dispersion of ε' and the relaxation loss peaks for ε'' in (Fig. 5.7) are based on the Debye equations[5.13].

$$\left.\begin{array}{l} \varepsilon'(\omega) = \dfrac{\varepsilon_0' - \varepsilon_\infty'}{1 + \omega^2 T^2} \\[4mm] \varepsilon''(\omega) = \left(\varepsilon_0' - \varepsilon_\infty'\right) \cdot \dfrac{\omega T}{1 + \omega^2 T^2} \end{array}\right\} \qquad (5.26)$$

Here ε'_∞ is the ε' value at frequencies higher than the dispersion region ($= n^2$ if atomic polarisation is negligible).

ε'_0 is the plateau value at frequencies below the dispersion region.

τ is the relaxation time. This derives from a single exponential time dependence for the d.c. step response:- $i \propto e^{-\frac{t}{\tau}}$

The ε'' maximum occurs where $\omega \tau = 1$ and equals $\frac{\varepsilon'_0 - \varepsilon'_\infty}{2}$ i.e. 1/2 of the dielectric increment for the process.

For dielectric liquids with small polar molecules $f_m = \frac{\omega_m}{2\pi}$ values lie in the frequency range 10^8 to 10^{11} Hz e.g. the peak for water at 20°C occurs at 1.7×10^{10} Hz [5.14]; and the width of the peak is very little greater than the Debye width. ($\Delta \log \omega = 1.14$ at 1/2 height).

In contrast, the width of the peak for (solid) polytetra-fluoroethylene in (Fig. 5.6) is 3 decades and that for polyethylene is about 5 decades. The latter is however known to be composite involving two relaxation mechanisms [5.15].

These are very low intensity loss peaks for near ideal materials. The magnitudes of the loss peaks and related dielectric increments, $\Delta \varepsilon'$ which are observed for highly polar solids are several orders greater. The peaks can occur anywhere in the spectrum from 10^{10} Hz downwards depending on the degree to which the motion of dipolar groups is hindered by neighbouring molecules in the solid matrix. For example in (Fig. 5.8) the room temperature results for the organic glass polymethylmethacrylate show an ε' increment of 2.3, an ε'' peak of 0.18 with a 1/2 width of 4 decades occurring around 10 Hz.

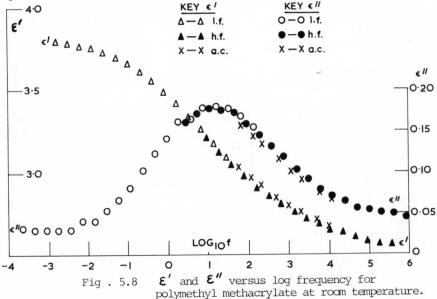

Fig . 5.8 ε' and ε'' versus log frequency for polymethyl methacrylate at room temperature.

These results were obtained by one of the author's colleagues using a Fourier Transform d.c. Step Response spectrometer [5.16] and published in review papers [5.17, 5.18]. The data effectively bridged the gap in the frequency coverage at room temperature in earlier published data reproduced in (Fig. 5.9[5.19]).

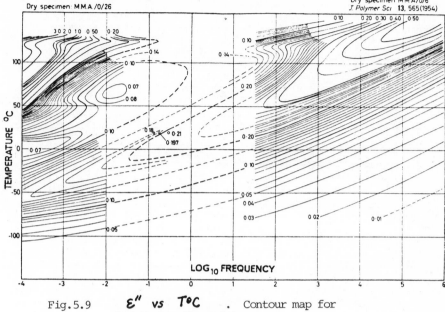

Fig.5.9 $\mathcal{E}''$ vs $T°C$. Contour map for polymethyl methacrylate.

The low frequency part of this contour map was obtained using the type of circuit given in (Figs. 5.1), with the sample contained in a servo-controlled dry air thermostat. The right hand side had been published already[5.20] in the form of separate $\mathcal{E}'$ and $\tan\delta$ contour maps.

(Fig. 5.9) illustrates the value of the contour plot as a means of separating characteristic features; attention is directed to the following points.

Firstly, the dominant relaxation loss peak moves from $\log f = -3.5$ at -50°C to $\log f$ = + 5.5 at + 140°C, increasing in magnitude progressively from 0.17 to 0.5. The ½ width increases as temperature falls, being about 7 decades at -50°C.

Secondly, at about 120°C there is a near constant temperature ridge discernable on both sides of the frequency gap. This is a minor feature electrically, but it correlates with the main glass/rubber transition region which exhibits a dramatic change of mechanical properties. Deutsch et al[5.20] argued that the 120° transition involves molecular rotation about the chain axis, but this motion does not involve a large dipole, whereas the dominant loss is attributable to motion of the carboxy methyl side group which does involve a large dipole. The high $\mathcal{E}''$ values at high frequencies and high temperatures results from the combination of the two features. The high $\mathcal{E}''$ values at low frequencies and high temperatures is attributable to conduction effects (which affect the d.c. step response results in a manner very similar to the results for PVC in (Fig. 5.3)).

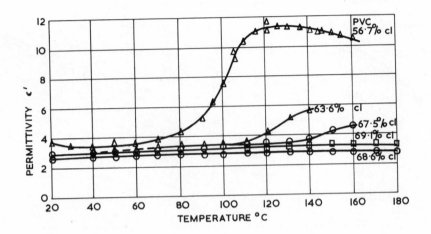

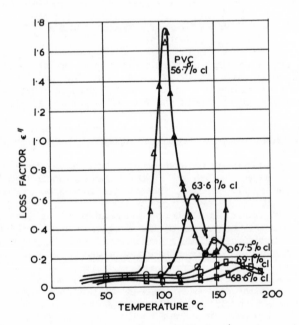

Fig.5.10 $\mathcal{E}'$ and $\mathcal{E}''$ vs temperature at 1000 Hz for chlorinated PVC's.

Many more examples of these patterns of behaviour have been published (5.18, 5.19). The first polymer material studies by the author in this comprehensive way was polyethylene terepthallate (terylene) (5.21).

If we consider again (Fig. 5.7), we realise that the phenomena in dipolar solids completely overlaps the region of the spectrum where interfacial polarisation appears. This in fact leads to serious difficulties of diagnosis when unknown materials are studied as it is not self evident from measurement results which class of mechanism is involved. Meakins (5.22) has treated this subject and also presents data on low frequency loss mechanism in inorganic compounds. His chapter on organic compounds is very full and detailed. The question - which type of mechanism - is usually resolved by study over a wide temperature and frequency range of related compounds with systematic variation of chemistry. For example the author was able to attribute the behaviour of polyvinyl chloride and a range of chlorinated PVC's to motion of dipoles about the chain axis (5.3), simply because the extra chlorine, which added additional C-Cl dipoles to the molecules had the effect of reducing the dielectric increment dramatically. This could occur only by a dipole balancing effect, which is expected from the nature of the chemistry. These results are given in (Fig. 5.10).

5.4 ELECTRETS AND FERROELECTRIC MATERIALS

So far we have discussed polarisation entirely as an effect of the applied electric field, the assumption being that dielectric materials relax back to a non-polarised equilibrium state when no field is present. I must now show that this very basic assumption is in most cases not strictly valid and in some is categorically not valid. The simplest approach to this is to discuss firstly electret and then ferroelectric substances.

If a sample of a good insulator is placed in contact with a metal electrode on one side, and the other is charged say by a high voltage corona discharge or by electron bombardment, then a strong electric field is set up between the free surface charge and its image charge in the metal. The material is polarised or 'poled' and this polarisation persists for a long time. Alternatively, if a dipolar molecular material is polarised with a high voltage at a temperature high enough to permit dipole motion - i.e. above the melting point for crystalline materials or the glass/rubber temperature for amorphous materials - and then cooled down in the presence of the field, the dipoles are frozen in with directions biased in the field direction. When the field is removed the orientation polarisation decays at a very slow rate and can thus persist for a long time. The proceedings of a symposium on electrets and dielectrics held in San Carlos, Brazil in 1975 (5.23) contains a wealth of references to the leading workers in this field.

In view of these mechanisms for production of permanent macroscopic polarisation of samples of material, it will not surprise you to learn that large quantities of good insulating materials are to a greater or lesser degree polarised adventitiously in the course of manufacture by contact charging etc. (5.24). Particularly is this the case for thin film material, and the author and his colleagues developed methods of studying these materials.

When electrets (adventitious or deliberate) are placed between electrodes and 'short circuited', the discharge current being measured using a circuit as (Fig. 5.1), as the temperature is raised at a steady rate, then 'thermally stimulated currents' can be recorded. This is one of the classical methods of studying electrets. In the course of this treatment the trapped charges and/or polarisation are released and the sample can be brought to a non-polarised state.

The distinguishing feature of ferroelectrics, which are crystalline substances, is that in certain temperature ranges the crystals polarise spontaneously. The minimum potential energy conditions, which determine the equilibrium positions of the various ions in the lattice, are satisfied with nett charge separations in certain directions. This tends to occur in domains and adjacent domains tend to polarise in opposite directions to maintain a zero nett polarisation. The effect of an electric field on such a system is much more dramatic than for the dielectrics discussed hitherto, and is characterised by a hysteresis loop for dielectric displacement D as a function of E (Equation 5.16), the relationship being extremely non-linear. The ratio $D/_E = \varepsilon$ can attain very high values (e.g. 10 000) at certain temperatures (the Curie temperatures). The Curie temperature can be varied by systematic ionic substitution as for example in the Ba TiO_3/Sr TiO_3 solid solution system. Basic work on ferroelectrics is done with single crystals. Commercially, polycrystalline ceramic versions are exploited as capacitor materials. Complex blends of materials with different Curie temperatures are used to achieve high permittivity values over a useful temperature range. (5.25, 5.26, 5.27, 5.28, 5.29).

REFERENCES

5.1 Strong, 1938, "Modern Physical Laboratory Practice". Blackie, Chapter 6.

5.2 Reddish, 1958, "Conduction and Polarisation processes in plasticised Polyvinylchloride Compounds". Society of Chemical Industry Symposium March 1958.

5.3 Reddish, 1966, "Dielectric studies of the transition regions for Polyvinylchloride and some chlorinated PVC's". J. Poly. Sci Part C 14 p.131, 1966.

5.4 Kouenhover and Baros, "A sensitive power factor bridge". Trans. Amer. I.E.E. 1932 51 p.202.

5.5 Hartshorn and Ward, "The Measurement of permittivity and power factor of dielectrics from 10^4 to 10^8 cycles per second". J.I.E.E. 1936 79 p.597, 1936.

5.6 Reddish, Bishop, Buckingham and Hyde, 1971. "Precise measurement of dielectric properties at radio frequencies". Proc I.E.E. 1971 118 No.1, pp 255.

5.7 Redheffer, "The measurement of dielectric constants", M.I.T. Radiation Lab. Services. The technique of microwave measurements" Mcgraw Hill, 1947.

5.8　Denbigh, "The polarisability of bonds", Trans. Far. Soc. 1940 36 p.336, 1940.

5.9　Thompson, "A course in chemical spectroscopy", Oxford, Clarendon Press, 1928.

5.10 Sutherland, "Infra red and Raman Spectra" Methuen Monograph, 1935.

5.11 Parker, Ford and Chambers, "The optical constants of pure fused quartz in the far infra red", Infra red Physics 1978 18. p.215.

5.12 Parker, M. and Chambers, "Determination of the complex dielectric response of solids in the far infra red from 4 to 300°K by dispersive Fourier Transform Spectroscopy" I.E.E. Conf. Publ. No. 177 1979.

5.13 Debye 1928, "Polar Molecules", English Edition 1945, Dover.

5.14 Hasted, "The dielectric properties of water", Progress in Dielectrics Vol. 3, Heywood, 1961.

5.15 Barrie, Buckingham and Reddish, W, "Dielectric properties of polythene for submarine telephone cables". Proc. I.E.E. 1966 113 No. 11, pp 7, 1849.

5.16 Hyde 1970, "A wide frequency range dielectric spectrometer", Proc I.E.E. 1970, 117 No. 9, pp 1891.

5.17 Mansel Davies, "Aspects of Recent Dielectric Studies". J. Chem. Soc., 1972.

5.18 Reddish, W. "Polymers for electrical insulation. A review of properties in relation to structure", Material Science Club Bulletin, 1975.

5.19 Reddish W. "Structure and dielectric properties of polymers" Pure and Applied Physics, Vol. 5, Butterworths, 1962, p 723.

5.20 Deutsch, Hoff and Reddish, W. Relation between structure of polymers and their dynamic mechanical and electrical properties", Pt. 1. Some alpha-substituted acrylic ester polymers. J. Poly Sci. 1954, XIII p.565.

5.21 Reddish, W. "The dielectric properties of polyethylene terephthalate (Terylene)" Trans Far. Soc. No. 330, 1950 46, pt. 6, pp 459.

5.22 Meakins, "Mechanisms of dielectric absorption in solids". Progress in Dielectrics Vol. 3, Heywood, 1961.

5.23 International Symposium on electrets and dielectrics. San Carlos, Brazil, 1975.

5.24 Blythe, 1975, "Adventitious Electrets", J. of Electrostatics 1. (1975) p. 101-110. Elsevier.

5.25 Jona and Shirane, "Ferroelectric Crystals", Pergamon Press, 1962.

5.26 Burfoot, "Ferroelectrics, An introduction to the Physical Principles", Van Nostrand, 1967.

5.27 Fatuzzo and Merz. "Ferroelectricity" North-Holland, 1967.

5.28 Plessner and West, 1960, "High permittivity ceramics for capacitors", Progress in Dielectrics, 1960 2 p.167.

Chapter 6

Applications of gaseous insulants

H. M. Ryan

6.1 INTRODUCTION

Atmospheric air is the most abundant dielectric material
which has played a vital role in providing a basic insulating
function in almost all electrical components and equipment.
However, because it has a comparatively low dielectric
strength, large electrical clearances are required in air for
high voltage applications such as overhead line bundle
conductor or tower design or open-type EHV switchyards. Air
clearances will be discussed in Section 6.2. Much research
activity has taken place during the past 30 years to develop
other gases with even better characteristics than air and some
of the recent developments with other gases will be discussed
in Section 6.3. As discussed in chapter I, the dielectric
strength of air etc. is a function of gas density and advantage
has been taken of this fact in transmission switchgear, where
high pressure gases have been used to provide insulation and
arc interruption characteristics in heavy duty interrupters.
The concept of high pressure insulation has recently been
extended into combined metal clad switchgear and connecting
cables, termed gas insulated systems (GIS). These will be
described in Section 6.4.
Since gas does not offer mechanical support for live
conductors, switches etc., it is used in combination with solid
insulation. The gas/solid interface region must be looked at
very carefully from an electrical design viewpoint - otherwise
design weaknesses could result in service with the possibility
of consequential breakdown or flashover of equipment. Even
when the very best design principles have been adopted, breakdowns can still occur in air/solid interfaces under adverse
environmental conditions such as pollution, fog or humidity.
These aspects will be dealt with in Chapters 10 and 15.
Whatever dielectric media is used, be it solid, liquid
or gas or combinations thereof, it is essential that careful
consideration be given to the electrostatic field design of
insulating components in order to produce compact, efficient,
economic and reliable designs of high voltage equipment which
will produce trouble-free life in service. To this end,
designers rely on analytical field analysis techniques to
model electrostatic aspects of specific designs and, when
appropriate, to optimise shape and disposition of components.
The general availability of digital field techniques during
the past 20 years has greatly assisted equipment designers.

An indication of the purpose, scope and recent applications of field analysis to EHV equipment designs is given in Section 6.5.

6.2   ATMOSPHERIC AIR CLEARANCES

6.2.1 Test areas

In dry, unpolluted areas, air clearances between live equipment and earth can be defined reliably. Legg (6.1) has described the minimum clearances necessary for high voltage test areas. The majority of high voltage tests on equipment are carried out indoors to avoid large variations in atmospheric conditions. When high voltage laboratories are constructed, the test voltages to be used are generally known and the physical size of the laboratory is determined by the dimensions of the voltage generating plant, the test objects and the clearances required to prevent flashover to the walls of the laboratory.

The flashover voltage of rod-rod and rod-plane gaps as a function of the gap spacing is illustrated in Fig. 6.1. From these graphs the dimensions of laboratories can be estimated. It will be noted that the lowest flashover voltages are experienced with positive switching surges and 50 Hz, applied to rod-plane gaps. For this reason, the performance of a rod-plane gap under positive switching surges may be usefully taken as a datum because all other combinations of gap geometry and voltage waveform produce higher flashover voltages. The sparkover of any gap is subject to statistical variation and some knowledge of this is required before clearances can be estimated for a laboratory. The breakdown voltages reproduced in Figs. 6.1 and 6.2 relate to 50% flashover levels, that is, if a large number of voltage tests at this level were applied to the gap, one half of these would be expected to cause flashover. Breakdown to the walls of a laboratory or test-chamber must be avoided; therefore greater clearances than these are required.

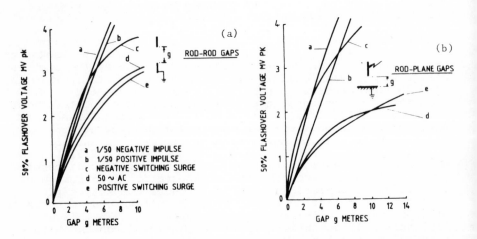

Fig. 6.1   50% Flashover voltage for rod-rod and rod-plane gaps
(Ref. 6.1)

In Fig. 6.2,the 50% flashover voltage is 1200 kV for this particular gap. The voltage corresponding to the 16% probability line (1090 kV) is one standard deviation (σ) less than V_{50} i.e. $V_{50} - \sigma$. The voltage corresponding to the 2% line is $V_{50} - 2\sigma$ (980 kV) and that to the 0.1% line is $V_{50} - 3\sigma$ (870 kV).

In practice, one takes the withstand voltage of the gap as 2 or 3 standard deviations less than the 50% flashover voltage, i.e. $V_{50} - 2\sigma$ or $V_{50} - 3\sigma$. The withstand voltage of the gap represented by Fig. 6.2 is therefore 980 or 870 kV. For 1/50 impulse waves, the standard deviation is of the order of 1 to 4%, the lower values occurring at the highest voltages, whereas for switching surges the standard deviation is of the order of 5 - 12% and increases with voltage. For 50 cycle voltages up to 1500 kV rms, the standard deviation is approximately:

1.5	-	4%		for rod-plane gaps
0.5	-	2.5%		for rod-rod gaps
1.0	-	4%		for dry insulators
1.5	-	6%		for wet insulators.

Having established the withstand voltage for various gaps - or conversely the gap required to withstand the various test voltages - the dimensions of a laboratory can be estimated. Even clearances based on $V_{50} - 3\sigma$ are often increased for two reasons. Firstly, the walls of the laboratory can influence the flashover voltage and invalidate the test. Secondly, clearances are required for radio interference or partial discharge measurements. During partial discharge tests, one may be looking for very small internal discharges (often as small as 1 pC) in the test object so that spurious discharges from the circuit (busbars, transformers, etc) must be an order smaller. If the clearances are too small, even though they are adequate to prevent flashover, then small discharges may flow to the walls of the laboratory thereby preventing accurate measurements. Test area dimensions are given in Table 6.1. For very high voltage (> 3 MV) test equipment, it may be uneconomic to build an enclosure with adequate clearances and the test equipment can be moved and used outdoors.

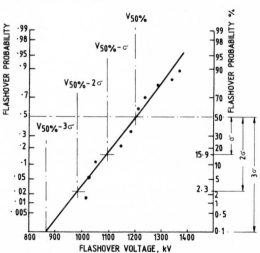

Fig. 6.2
Illustrating probability
of flashover   (Ref. 6.1)

Table 6.1. Approximate dimensions of test hall for 50 Hz
3 phase tests (Ref. 6.1)

Transformer voltage rating	H.T. to earth clearance, X				Nominal test room dimensions in feet			Diameter of discharge free busbars	
	Minimum practical		Discharge free at full voltage						
(kV rms)	(ft)	(m)	(ft)	(m)	L	W	H	(in)	(cm)
100	1.3	0.4	2	0.6	15	10	10	2	5
250	3	0.9	5	1.5	20	15	15	4	10
500	7	2.1	10	3	40	25	20	8	20
800	13	3.9	20	6	50	40	35	12	30
1000	19	5.8	25	7.6	75	55	50	15	40

L = length;   W = width;   H = height

6.2.2 Sphere gaps

A convenient use of air gaps is as a calibration for high
voltages.   It is useful in that no voltage dividers or
electronic equipment is required (which can malfunction), but
suffers from lack of accuracy, $\pm$ 3%, (covers the whole range of
voltage).   Spark gaps are recognised in BS: 358, from which
Fig. 6.3 can be derived showing voltage vs effective gap
($\propto = \eta g$) for different sphere diameters.   The sphere gap is
accurate and reproducible over its range, which extends to
gaps equal to 0.5 x sphere diameter, after which it loses
accuracy due to field distortion.   It is also necessary to
take account of humidity and pressure, both of which effect
the breakdown voltage.   The sphere gap records peak voltage
and can therefore be used for dc, ac or impulse voltages with
only slight variations for positive or negative pulses.   For

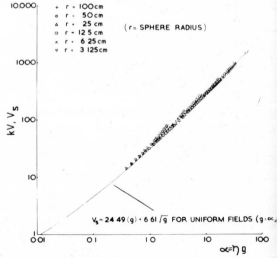

Fig. 6.3
Correlation of BS 358:1960
Sphere gap data
                (After Ryan)

very fast impulses (<10 μS), it is necessary to irradiate the gap with a radioactive source in one sphere to provide the necessary initiating electrons. Sphere gaps are no longer favoured in modern UHV laboratories, precision voltage dividers are now used - in preference - for which accurate calibrations are available.

6.2.3 Spark gaps

Spark gaps are used for protection of equipment against high voltage surges, the advantage being that the dielectric is recoverable. Spark gaps range from the common GPO arrester, having low pressure N_2 in a ceramic body, with voltage ratings of 150 - 1000V and current ratings of 103 amp.sec, to low pressure gas or vacuum "bottles" with voltage ratings of 1 to 50 kV and current ratings of 50 kA for e.g. railway equipment protection. The simplest and most common spark gap is the rod-gap, used on outdoor equipment such as transformers, substations, bushings. Its characteristics are well documented (e.g. Fig. 6.4); its simplicity is offset by a variable breakdown voltage with respect to time and polarity. Special designs with improved characteristics have been developed recently. For equipment protected by a rod gap, it is important to know its time/voltage characteristic relative

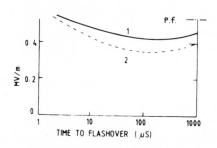

Typical relationships for critical flashover voltage per metre as a function of time to flashover (3 m gap): 1. rod-rod gap; 2. conductor-plane gap; p.f. power frequency CFO

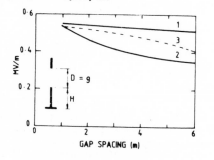

Typical relationships for flashover voltage per metre as a function of gap spacing: 1. 1.2/50 μs impulse; 2. 200/2000 μs impulse (rod-rod, H/D = 1.0, positive dry); 3. power frequency

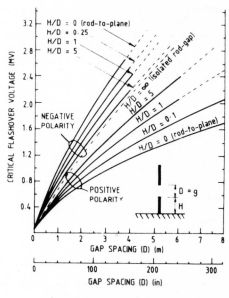

(Wave fronts of 100 - 200 μs)
(Vapour pressure ≈ 12.5 mean Hg)
Switching surge flashover strength of rod-rod and rod-plane gaps (courtesy Edison Electric Institute)

Fig. 6.4   Rod-gap characteristics (Ref. 6.2)

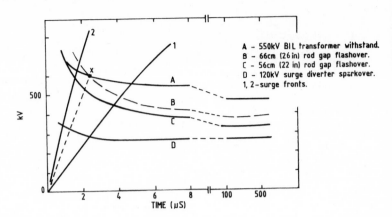

Fig. 6.5 Transformer protection by rod-gap and surge arrester
(Ref. 6.2)

to that of the equipment being protected to ensure that the
rod-gap operates first, under all surge conditions.   This
is known as insulation co-ordination (Fig. 6.5).

The role of rod-gaps and surge-arresters, widely used as
overvoltage protective devices, is summarised elsewhere (6.2).
Voltage-time 'withstand' characteristics exist for different
types of insulation when subjected to various types of overvoltage — the various time regions refer to particular types
or shapes of test voltage.   The 'withstand' levels used for
self-restoring insulation are at a suitably chosen level below
the V_{50} vs time characteristic (a selected multiple of σ below),
while characteristics for non-self-restoring insulation are
obtained from withstand characteristics corresponding to the
specified test voltages for the various categories of test
voltage.   The example reproduced in Fig. 6.5 illustrates that
protection can be achieved in any time region in which the
protective characteristic lies below the withstand characteristics of the insulation concerned.   The 120 kV surge arrester
(curve D) protects the transformer over the entire time-range
whereas the 26 in. rod-gap (B) protects the transformer only
against surges with front slopes less than OX. Steeper surges
would cause the insulation to breakdown before the rod-gap
could operate.   To achieve reliable protection, comprehensive
test data must exist both for the protective devices and the
insulation components of any system.

6.2.4 Overhead lines and conductor bundles

Various classifications of dielectric stress may be
encountered during the operation of any equipment ranging from
(i) sustained normal power frequency voltages, (ii) temporary
overvoltages, (iii) switching overvoltages to (iv) lightning
overvoltages.   These are represented in the laboratory. for
test purposes, by standard test waves : power frequency,
switching impulse and lightning voltages.   The strength of
external insulation is dependent on geometric factors, on air
density, humidity (appropriate correction factors are
available), precipitation and contamination (due to natural or
industrial pollution).

Before effective insulation coordination can be achieved, a large amount of experimental test data is required on arrangements such as: support insulators (various shapes), rod-rod gaps (for protective purposes) rod-plane, ring-plane gaps etc. which approximate to practical conditions (e.g. air clearances between phases, phase-earth clearances) and insulator strings for transmission lines (vertical, horizontal vee-strings, disc, longrod, fog-type etc.). A variety of typical characteristics for various arrangements, have been summarised and presented by Diesendorf (6.2) and examples are reproduced in Figs. 6.4 and 6.6.

The mean breakdown stress (i.e. V/g) in large practical gaps (Fig. 6.4) is typically 500 V/mm, somewhat reduced from the 2 kV/mm in sphere gaps (Fig. 6.3). This can reduce even further in wet weather, such that breakdown of air under lightning conditions can occur at stresses as low as 20 V/mm.

In overhead line design (see 6.2), it is necessary to design conductor bundles such that the stress at the surface of the bundle is less than that to initiate corona, otherwise radio noise would be generated and losses and corrosion would increase. An indication of surface stress levels of various UHV conductor bundle designs are given in Fig. 6.7. Special problems to be overcome before effective overvoltage protection can be achieved in practice are considered elsewhere (6.2).

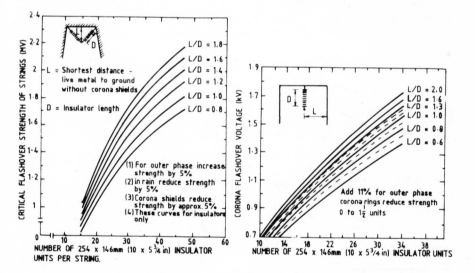

Curves for dry positive-polarity switching-surge flashover strength of V-string insulators

Curves for dry positive-polarity switching-surge flashover along tangent insulator strings in centre phase of steel towers

Fig.6.6 Switching surge characteristics of string insulators*

* EHV Transmission Line Reference Book Edison Electric Inst., New York 1968

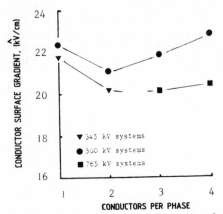

Fig.6.7
Typical surface gradients on
overhead transmission line
conductors*

6.3   OTHER GASES

The availability and characteristics of air makes it the
most used gas insulant.   In enclosed equipment, e.g. heavy
duty gas blast circuit-breakers, by using air at high gas
pressures, the dielectric performance etc. could be greatly
enhanced.   However, because of limitations, gases other than
air have been considered. The most popular of these has been
sulphur hexaflouride gas (SF_6) which has dielectric strength
about twice as good as air and also offers excellent thermal
and arc interruption characteristics. Despite this, the search
continues for other gases with improved characteristics (6.4).
There has been considerable research activity (6.5)to find
alternative gases and gas mixtures (see Figs.6.8, 6.9), which
may have comparable breakdown strengths as SF_6 but can offer
technical and economic advantages.   Binary mixtures with in-
expensive gases like N_2, air, CO_2 and N_2O have been under
continuous investigation in order to arrive at efficient and
economical mixtures and obtain a better understanding of the
dielectric processes involved.   The mixture of SF_6 and N_2 is
the only mixture to achieve commercial application in switchgear to satisfy low ambient temperature conditions.

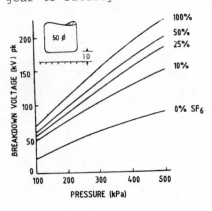

Fig. 6.8
Uniform field breakdown
voltages for SF_6-air
mixtures

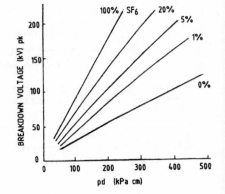

Fig. 6.9
Paschen Curves for SF_6-N_2
mixtures at 20°C (50 Hz)

Ternary gaseous dielectrics are also being investigated, such as N_2 + SF_6 or CHF_3 + SF_6, in conjunction with perfluoro-carbons. Fluorocarbons, when mixed with SF_6, may exhibit interesting synergistic effects which are at present unexplained. A variety of gaseous insulating systems is being investigated, comprising multi-component gas mixtures carefully selected on the basis of physicochemical knowledge, especially on the interactions of low energy electrons with atoms and molecules. Much work in this field is being actively pursued by a group led by L.G. Christophorou at the Oak Ridge National Laboratory, Knoxville, Tennessee, USA.

6.4   SWITCHGEAR AND GIS

6.4.1   Introduction

High voltage switchgear which forms an integral part of any substation (i.e. switching station), is essentially a combination of switching and measuring devices. The switches (circuit-breakers),connect and disconnect the circuits and the measuring devices (instrument transformers), monitor the system and detect faults. Particular care must be given to the selection of switchgear since security of power supply is dependent on their reliability. The reliability of any circuit-breaker depends on: insulation security, circuit-breaking capability, mechanical design and current carrying capacity. Considering modern switchgear practice, circuit-breakers can be broadly classified according to the insulating medium used for arc-extinction:

o bulk-oil   o air-break   o SF_6 gas
o small oil-volume   o air-blast   o vacuum.

Several types exist for indoor or outdoor service conditions. There are established relationships between types of switch-gear selected and the design of the system. Briefly, until a few years ago;

- for voltages up to 11 kV, most circuit-breakers were of the oil-break type, or of the air-break type, using air at atmospheric pressure.

- from 11 to 66 kV, oil circuit-breakers were mainly used while, at 132 and 275 kV, the market was shared by oil interrupters (both bulk-oil and small oil-volume) and gas blast circuit-breakers.

Recent developments have seen the increased demand for vacuum and SF_6 circuit-breakers for distribution systems and the increased use of SF_6 circuit-breakers for system voltages above 132, up to and beyond 400 kV. However, modular-small-oil volume designs have still proved popular for all but the highest ratings (6.6).

Historically, it is worth noting that in the nineteen-sixties, the second generation air-blast transmission circuit-breakers, for 420 kV, had six breaks per phase. The predict-ion that such interrupters would develop to higher rating was not fulfilled and the recent trend has been towards compact SF_6 switchgear. By incremental research, it has been possible to develop SF_6 dead-tank circuit-breakers for 420 kV, 63 kA rating, with only two-breaks per phase.

6.4.2   Arc Extinction Media

The basic construction of any circuit-breaker entails the separation of contacts in an insulating 'fluid'.   The insulating 'fluid' which fills the circuit-breaker chamber must fulfil a dual function.   Firstly, it must extinguish the arc drawn between the contacts when the circuit-breaker opens and secondly, it must provide adequate - and totally reliable - electrical insulation between the contacts and from each contact to earth.

Selection of arc extinction 'fluid' is dependent on the rating and type of circuit-breaker.   The insulating media commonly used for circuit-breakers are:

o   air at atmospheric pressure
o   oil (which produces hydrogen for arc-extinction)
o   compressed air (at pressure ≈ 7 MPa)
o   sulphur hexafluoride (SF_6) (at pressure < 0.85 MPa)
o   vacuum.

The principle of circuit interruption techniques in circuit-breakers, either of single or multiple break design, has already been covered elsewhere (6.6, 6.7).   However, it is appropriate to comment briefly on interruption techniques, relevant to SF_6 circuit-breakers, to illustrate basic principles.   The discharge products of SF_6 are discussed in Chapter 10.

Table 6.2 provides a convenient summary of single or double pressure SF_6 interrupter processes together with a brief explanation of live-tank and dead-tank design philosophies.

6.4.3   General dielectric considerations

Because of the widespread use of SF_6 gas in gas insulated switchgear (GIS) it is appropriate to touch briefly on some general points relating to the dielectric performance of gas-gaps and gas/solid insulation interfaces in SF_6 under clean and contaminated conditions.   The literature on this subject is vast and extensive bibliographies exist (6.8 - 6.10) listing some important papers which can be arbitrarily grouped in topic areas; particulate contamination and detection, diagnostic techniques, operating experience, breakdown studies, internal-arcing, spacer experience and dielectric discharge performance.

6.4.3.1   Dielectric withstand capabilities

Gas-Gap Data: Typical limits of lightning and switching impulse withstand gradient performance for a family of large practical GIS gas-gap type configurations are illustrated in Fig. 6.10 for SF_6 pressures in the range $(0.1 < p < 0.6)$ MPa. Experimental determination of critical breakdown (E_{50}) and highest withstand (E_W) gradient values were obtained using similar test techniques to those adopted previously (6.11). It should be emphasised that these curves present typical design type gradient relationships and encompass the results obtained for a large family of coaxial cylinder and perturbed electrode configurations, for gas-gaps in the range $50 < g < 180$mm. Corresponding withstand curve for 50 Hz conditions is also shown.

Table 6.2 SF₆ Circuit-breaker concepts

Single or double-pressure

The flow of SF₆ gas necessary for circuit-breaking may be produced in either of two ways. In both cases, however, the actual process of current interruption is similar, i.e. the arc is established through a nozzle by the separation of the contacts and/or gas flow and is subjected to an 'axial' gas blast which abstracts energy from the arc resulting in extinction at a current zero.

The gas flow may be achieved by the use of a two-pressure system (Fig. 1) whereby the operation of a blast valve allows gas to flow from a high pressure reservoir through a nozzle into a low pressure reservoir. The alternative is a single-pressure system (Fig. 2) where compression of the gas is caused by movement of a cylinder over a fixed piston (or vice-versa). In this system, commonly called the 'puffer' (Fig. 3), the moving contact system and cylinder are usually joined together so that the movement of only a single component is necessary for arc initiation and interruption.

The single-pressure system is inherently less complicated than the double-pressure, which requires two separate gas reservoirs with associated seals, a compressor and gas-handling system, and heaters to prevent liquefaction at low temperatures. Furthermore, in the double-pressure system there is the necessity of synchronizing the blast-valve and the contact driving systems.

Very high unit interrupting ratings can now be achieved with both systems, though the high-performance single-pressure system requires a more powerful operating mechanism to ensure a sufficiently high velocity for the contact/cylinder arrangement. This, however, is easily obtainable with either pneumatic or hydraulic power units, and at the lower performance levels, spring opening mechanisms also offer a practical alternative.

Initially the performance limitations of the single-pressure system meant that earlier types of circuit-breaker incorporated double-pressure interrupters; but the continuing development of the single-pressure puffer interrupter has led to designs which are capable of the highest ratings. It seems likely, therefore, that such interrupter systems will be the basis of most future e.h.v. circuit-breakers. They will also find wider application in the high-voltage distribution field.

In any axial gas blast interrupter, the unit performance may be significantly improved by using a dual-blast construction in which the arc is drawn through two nozzles. In this construction, the gas flow is towards both arc roots and metal vapour from the contacts is not blown into the arc column. In the partial duo-blast construction (Fig. 4) one of these nozzles has a smaller diameter, which reduces the amount of gas passed through the nozzles and hence also the mechanical energy input without a significant reduction in performance compared to the full duo-nozzle construction. The Reyrolle puffer interrupters incorporate partial duo-blast constructions.

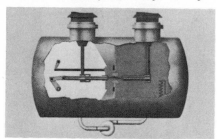

Fig. 1 Double-pressure system

Fig. 2 Single-pressure system

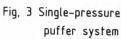

Fig, 3 Single-pressure puffer system

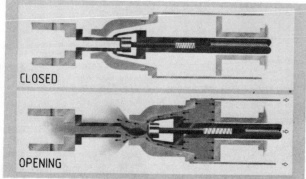

CLOSED

OPENING

Table 6.2   SF₆ Circuit-breaker concepts (cont.)

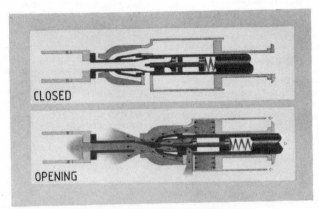

Fig. 4   Partial duo-blast puffer system

Live or dead tank

The SF₆ interrupter can be incorporated in either live-tank or dead-tank circuit-breakers (Fig. 5). The type chosen will depend on economics and/or the type of application; for example, the dead-tank construction (in which all interrupters are enclosed within an earthed pressure-vessel) is essential for use in complete metalclad installations, although the same breaker may be used with terminal bushings in open-type layouts. In assessing relative economics of live- and dead-tank constructions, the cost of any associated current transformers should be considered. Accommodation for current transformers is, of course, integral in the dead-tank circuit-breaker, but separate post-type units are usually necessary with the live-tank design for outdoor installations. With indoor installations the current transformers may be accommodated in the through-wall bushings or cable-sealing ends.

Generally it might be expected that at the lower end of the voltage/current scale where only one or two series breaks per phase are required, i.e. up to 420 kV the live-tank construction is probably more economical; but at higher voltages with three or four (or more) series breaks, the dead-tank construction tends to have an economic advantage because of the reduced amount of external insulation.

It is worthwhile noting that dead-tank circuit-breakers are inherently more suitable for areas subject to earthquakes.

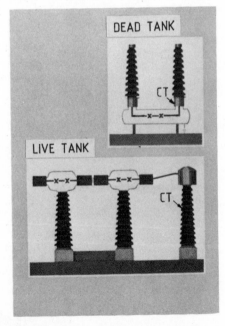

(Courtesy NEI Reyrolle Ltd)

Fig. 5 Live and dead-tank circuit-breakers

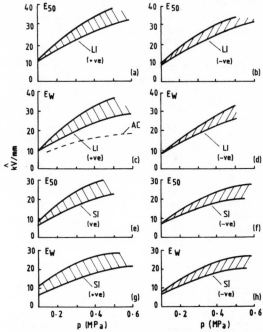

Note:

Results relate to several large systems representative of GIS electrode shapes, dispositions etc., covering gas-gaps in the range (50 < g < 180) mm

Fig. 6.10   Typical limits of 50% breakdown gradient (E_{50}) and critical withstand gradient (E_W) on SF_6 pressure for large coaxial and perturbed cylindrical electrode systems under clean conditions (6.8).

Curves

a.b.c.d. – Lightning impulse waveshape (1.2/50µs)
e.f.g.h. – Switching impulse waveshape (250/250µs)
– – – – – Also shown in curve (c), lower limiting 50 Hz withstand characteristics (E_W)

Extensive data exists for SF_6 gas-gaps under clean and contaminated conditions (e.g. see Ref. (6.8 – 6.13). It is now established that the presence of particulate contamination of lengths 2 – 20 mm can reduce the dielectric withstand capabilities of practical gaps by varying amounts up to typically 30, 40 & 70% for lightning impulse, switching impulse and power frequency conditions respectively, at working SF_6 pressure.

Barrier Performance Data: Chapter 10 will consider GIS support barriers.   It should be noted that the withstand characteristics of SF_6/support barriers under clean conditions are dependent on resin formulation, insulation shape and on the disposition of stress relieving fitting, inserts etc.   Typical withstand gradient levels of 11.6, 8.7 and 6.6 kV pk/mm can be achieved under lightning, switching impulse and 50 Hz short term voltage conditions respectively (6.8).

The presence of particulate contamination can reduce the 50 Hz withstand capability of cast-resin support barriers by varying amounts (e.g. up to < 30%) depending on particulate size and disposition.   For the most onerous dispositions of

cast-resin support barriers, the percentage lowering of withstand performance under impulse conditions tends to be much
less than that experienced for 50 Hz test conditions, for
comparable levels of contamination .

6.4.3.2 GIS equipment: Power frequency (V-t) characteristics

An understanding of voltage gradients that can be safely
sustained in service in GIS emerges from extensive laboratory
studies, relating specifically to long-time 50 Hz test
conditions. The ratio E/p is a convenient measure of the
stress applied to GIS equipment in service. Typical normalised working stress (E/p) levels are 7 MV/MPa and 3 MV/MPa
for SF_6 switchgear and also instrument transformers, (Fig.6.11)
manufactured by the author's company for 300 and 420 kV
systems. Many of these units have been in service for
periods up to 15 years and excellent service reliability has
been demonstrated. It is readily apparent that the maximum
service stress levels for gas-gaps in SF_6 insulated switchgear are relatively low, i.e. < 4kVrms/mm, compared to attainable withstand characteristics, (see curves, Fig. 6.10).

The working stress levels shown in this figure can be
considered to be generally representative of GIS.   By
restricting these gradient levels well below the limiting
values, deduced from long-term laboratory studies, manufacturers hope to ensure the long-term dielectric integrity of
equipment - provided of course normal quality and service
procedures have been maintained. Obviously, the question of
component cleanliness achieved during construction in the
factory and during site assembly is an important factor in
the design of GIS; critical areas of metalclad designs have
been identified which merit special manufacturing, testing
and assembly controls.

Several recent papers have considered the power frequency
(V-t) characteristics of model gas-gaps and gas/insulator
arrangements, (e.g. Ref. 6.13, 6.14) for large electrode
systems under 'gross' contamination conditions with varying
gas pressures.

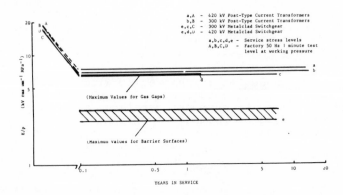

Fig. 6.11 Typical normalised 50 Hz service stress levels for GIS
equipment.

6.4.4 Performance under contaminated conditions

The achievable design stress and reliability of GIS apparatus under normal power frequency service conditions is crucially effected by particulate contamination. Particles in the gas space and on the insulator surface can significantly lower the dielectric strength of the system. In SF_6, the breakdown voltage for a relatively long particle, fixed in contact with the conductor, is considerably higher than that for a number of free conducting particles, over a limited range of pressures. Under power-frequency voltage excitation, free conducting particles tend to bounce along the bottom of the enclosure or across the surface of insulators. The amplitude of individual bounces depends on the particle size and shape, the potential on the system and other random parameters. Particle initiated breakdown can occur at voltages considerably lower than those required for breakdown due to the roughness of the electrode surfaces and is, in general, lower for negative impulse polarity voltages (6.12). During the past few years, switchgear and CGIT manufacturers - in support of GIS, equipment developments - have undertaken comprehensive evaluation of the dielectric performance characteristics of practical gas-gap arrangements and support insulation configurations in SF_6, under both clean and contaminated conditions. Such studies can involve comprehensive laboratory tests to determine, and quantify, the effects of particulate contamination size on the breakdown voltage in SF_6, under conditions representative of both gas insulated 'back-parts', circuit-breakers, disconnect switches, etc, for widely varying experimental conditions.

Numerous inter-related factors can influence the degree to which the presence of particles can lower the dielectric withstand capabilities of GIS under normal service conditions. These depend on:

- Length and diameter of particles present
- Whether particles are metallic or non-metallic
- Quantity and nature of contaminant material (density etc)
- Position of particles relative to electric field and also to various GIS components
- Actual design and physical disposition of GIS components (e.g. whether circuit-breaker, backparts etc. are mounted horizontally or vertically)
- Type of particle movement, when electrostatic forces exceed those of gravity
- Working gas pressure of SF_6 or gas mixtures
- Effectiveness of particle 'collection' or 'trapping' techniques.

Much valuable work has been done to develop suitable commercial particle 'traps' to overcome the potential problems of particle-initiated breakdown in GIS, or CGIT equipment, and, while it is widely recognised that such techniques can be very useful, especially if 'voltage conditioning' site testing techniques are also employed, to encourage the movement of particles into traps - there is still no universal agreement regarding the total effectiveness of such devices under all service situations.

6.4.5 Vacuum switches

Vacuum, at better than 10^{-6} torr, has an electric strength of typically 30 kV/mm; under arcing conditions, gas for ionisation is provided by molten metal at the arc root of the electrodes and vacuum breakdown is very dependent on electrode conditions.   Arc interruption in vacuum circuit-breakers (VCB) is therefore achieved (Ref. Reece 6.6), by cooling the arc root quickly to suppress the hot spot - this is achieved by rotating the arc root rapidly under its own magnetic field and by using electrode materials of high boiling point and good thermal conductivity.   The high electric strength of vacuum ensures that once the arc is suppressed, usually at the first current zero, no re-ignition occurs - dielectric recovery is achieved within a few µs.

The contacts are housed in a sealed glass or ceramic bottle (Fig. 6.12) with a moving metal bellows and are maintenance-free.   Excellent service performance has been demonstrated for fault currents of < 40 kA and 33 kV operation. VCB are ideal for use where many breaker operations are required, e.g. railways, arc furnaces.   The impulse level ($\approx$150 kV) on the small gap (10 mm) limits the working voltage of a single bottle, but bottles have been stacked in series to provide 132 kV breakers (Ref. 6.6).   The major use of VCB is in 11, 33 or 66 kV applications for distribution use where an oil-free maintenance-free, breaker is required.

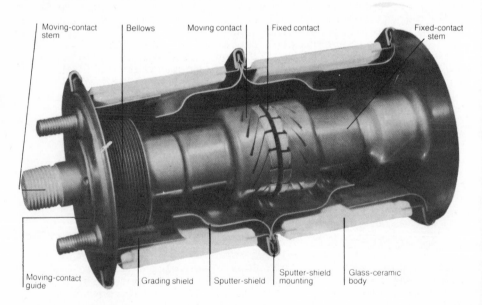

Fig. 6.12   Section view of vacuum interrupter
(Courtesy Vacuum Interrupters Ltd)

6.5   SYSTEM MODELLING

6.5.1 Field analysis techniques

The electrostatic fields of high voltage equipment are satisfied by the well known Laplacian equation which takes the form

$$\nabla^2\emptyset = \frac{\partial^2\emptyset}{\partial x^2} + \frac{\partial^2\emptyset}{\partial y^2} + \frac{\partial^2\emptyset}{\partial z^2} = 0$$

where $\emptyset$ is the potential at any point in the Cartesian co-ordinate system x, y, z.   The field distribution in any design is dependent on the shape, size and disposition of the electrodes and insulation and also, in general, on the permittivities of the insulating materials used.

One criterion which can be applied to the evaluation of the electrostatic design of high voltage equipment is the ratio of the average to maximum field.   This parameter, termed the utilisation factor η, is considered below, together with details of a related parameter termed the normalised effective electrode separation.   Brief reference is made in section 6.5.1.3 to a simple approximate 2-dimensional method which can often be used to estimate maximum voltage gradients for complicated arrangements which cannot yet be solved directly by available numerical methods.

Analytical solutions of the Laplace equation can only be obtained for relatively simple electrode systems, where the conducting surfaces (electrodes) are cylinders, spheres, spheroids or other surfaces conforming to equipotentials surrounding some simple charge distribution.   Generally, the multiplicity of boundary conditions for the complicated contours encountered in high-voltage equipment means that analytical solutions of the potential are not possible.   Because of this difficulty, several approximation methods have been investigated, the more important of these being (i) analogue methods and (ii) numerical methods (6.15).

6.5.1.1 Utilisation factor approach

In the evaluation of the electrostatic design of high voltage equipment, an important consideration is the effectiveness with which the available space has been used.   For the region which encompasses the minimum distance between conductors, the ratio of average to maximum electric stress (E_{av}/E_{max}) is a useful criterion.   This ratio, termed the utilisation factor η, measures the inferiority of the field system in comparison with that between infinite plane parallel electrodes (6.16) where $\eta = 1$.   In the literature, the reciprocal term $1/\eta$ called the field factor is sometimes preferred.   Tables of η, exist (6.16) for several standard electrode systems for wide ranges of so-called geometric characteristics p and q where $p = (r + g)/r$ and $q = R/r$. Subscripts 2 and 3 respectively are used to denote 2 and 3 dimensional systems.   Appropriate formulas are given elsewhere, use being made of existing solutions of Laplace's equation (6.15 - 6.22).   These calculations assume constant permittivity, but techniques can be extended to multi dielectric problems.

Some calculated values of η are also present in graphical form in Fig. 6.13. It is immediately evident that η for each geometry tends to reduce as the geometric characteristic $p = (r+g)/r$ increases. This can be explained by the increased divergence of the field as p increases. Furthermore, it is noted that for a particular value of p, η for a 3-dimensional geometry is lower than that for the corresponding 2-dimensional geometry (e.g. compare cylinder-plane and sphere-plane, curves 2 and 5, Fig. 6.13). Once again, this may be attributed to the greater field divergence in the 3-dimensional arrangements. Brief mention will be made later to the simple relationship existing between η for corresponding 2 and 3 dimensional systems.

To summarise, utilisation factors, or field factors, may be calculated for many practical arrangements either as a result of precise analysis for simple geometries or, for more difficult configurations, by approximate numerical or analogue techniques together with appropriate difference equations discussed elsewhere (6.17).

6.5.1.2 Efficiency factor concept

For certain design problems, conditions of fixed axial distance must be complied with. The utilisation factor has the limitation that it does not take into account the space necessary to accommodate the electrode geometry. To allow for this, η can be multiplied by the electrode separation g. The product of these terms gives the 'effective electrode separation' $\propto$ of the system. If, now, $\propto$ is divided by some measure of the total dimension available, a new quantity λ is

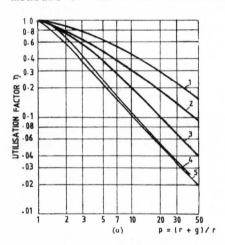
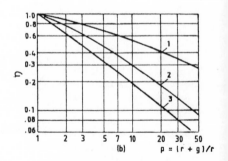

curves 1 cylinder – cylinder
      2 cylinder – plane
      3 sphere – sphere
         (symmetrical supply)
      4 sphere – sphere
         (unsymmetrical supply)
      5 sphere – plane

curves 1 hyperbolic cylinders
      2 hyperboloids
         (points)
      3 hyperboloid
         (point) – plane

Fig. 6.13 Dependence of η upon geometric characteristic p for a few simple geometries

obtained, which is really the normalised effective electrode separation of the system.   The efficiency factor concept (λ) is sometimes useful when designing high voltage equipment, particularly when conditions of fixed axial distance d exist (6.16).   For a particular operating voltage V, it is possible to keep the maximum stress at the electrode surface to a minimum by selecting the best size, shape and disposition of electrodes for conditions of constant d (see Fig. 6.14). Under these conditions $\lambda = \hat{\lambda}$ and $E_m = V/\hat{\lambda}d$.

All the results referred to above assume a single value for the dielectric constant.   If two or more dielectrics are being considered, then the concepts of utilisation factor and efficiency factor could be applied to each dielectric separately.   These techniques are now widely used in GIS design.

6.5.1.3 Approximate 2/3 dimensional concept

The electrode arrangements considered above contained at least one axis of symmetry.   In many practical arrangements however such symmetry does not exist.   In the absence of symmetry, the problem of making accurate pronouncements upon maximum field strengths for practical 3-dimensional systems often becomes complicated since it is sometimes either exceedingly difficult or even impossible to make single realistic numerical or analogue model representations without introducing appreciable errors.

It would be useful if, for example, one could obtain even an approximate solution for a particular practical 3-dimensional representations (having conductors similar to axial and radial sections of the original 3-dimensional conducting surfaces), which, in general, can be more easily solved. Boag and later Ryan investigated the possibilities of this simple approach and have analysed numerous simple configurations (Fig. 6.15).   Ryan has produced useful correction curves which can be used with acceptable accuracy.

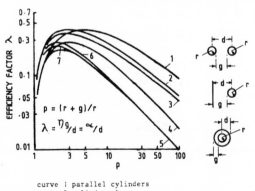

curve 1 parallel cylinders
curve 2 cylinder-plane
curve 3 concentric cylinders
curve 4 sphere-sphere (symmetrical supply)
curve 5 sphere-sphere (unsymmetrical supply)
curve 6 sphere-plane
curve 7 concentric spheres

Fig. 6.14
Dependence of λ upon p for several standard geometries

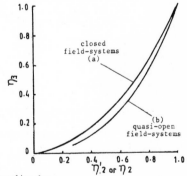

a applies for
  concentric cylinders and concentric spheres
  cylinder-plane and sphere-plane
  hyperbolic cylinder-plane and hyperboloid plane
  parabolic cylinders and paraboloids of revolution

b applies for systems shown in Fig. 6.17

Fig. 6.15   Correction curves
(Ref. 6.18)

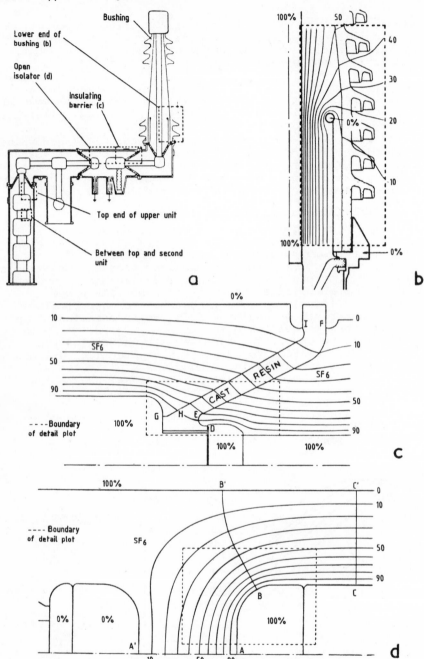

Fig.6.16 Computer field study of 300 kV SF$_6$ metalclad switchgear
(Ref. 6.22)
   a   part of single phase layout
   b   field plot at lower end of bushing
   c   field plot of barrier supporting isolator
   d   field plot in vicinity of open isolator

6.5.1.4 Numerical methods

Numerical methods of solution, which express the Laplacian equation in finite terms, provide a powerful means of calculating the electric fields of practical arrangements. A great deal of published literature exists relating to this subject.   For example:

o In contrast to other workers who have used relaxation techniques, Galloway et al (6.15, 6.22) have developed a computer program to solve the Laplace equation in 2-dimensions and 3-dimensions with one axis of symmetry by an exact non-iterative method. This method was selected after various techniques for solving the resultant set of simultaneous equations had been studied. The main details of this program etc. are given in Ref. 6.15, 6.22).

o Binns and Randal (6.23) have described an over-relaxation method and give details of various accelerated finite difference formulas used. In this investigation, potential gradients were calculated around a spherical high voltage electrode separated from an earthed plane of recessed dielectric slab. The potential gradient-transitions were determined and analysed at the point where the surface of the recessed dielectric slab meets the sphere surface.

o Storey and Billings (6.24) have described a successive over-relaxation method suitable for determining axially symmetric field distributions. They also discuss a method for the determination of the 3-dimensional electric field distribution in a curved bushing.

A detailed discussion of relaxation methods, difference formulas, accelerating factors etc. is outside the scope of this chapter.

6.5.1.5 Insulating spacer design

With the widespread interest in gas insulated cables, busbars etc., considerable attention has been given to the design of spacer shapes for such equipment. The examples summarised in Fig. 6.16 illustrate the extensive use made of analytical field techniques, during early insulation development work on a EHV SF_6 metalclad switchgear installation.

6.5.2 Prediction breakdown voltages

6.5.2.1 Empirical approach

In 1961, Ryan first investigated the feasibility of using simple estimation methods of predicting the minimum breakdown voltage levels of non-uniform field configurations in gas insulated equipment. Initially, the method suggested by Schwaiger was considered and later extended (6.16 - 6.22) to more complex electrode configurations by incorporating simple perturbation principles. The method does not consider breakdown mechanisms but is based on a simple discharge-law concept (6.21).

By ignoring space charge effects, the breakdown voltage V_s is given by the relationship

$$V_s = E.\eta.g. \qquad \ldots \qquad \ldots \qquad \ldots \qquad (6.1)$$

where E and g are the appropriate breakdown gradient and gap dimension respectively and η, derived from Laplacian field analysis, is the utilisation factor (ratio of average to maximum voltage gradient (E_{av}/E_m). Initially, standard electrode geometries amenable to precise electrostatic field solution were considered and extensive tabulated field data has been published (6.16 - 6.17). Results of early investig- ations (e.g. Fig. 6.17) firmly established the usefulness of this simple concept for estimating minimum breakdown voltage levels.

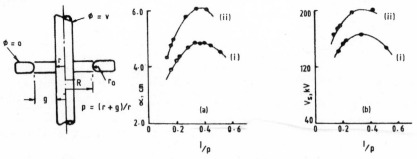

Fig. 6.17   Dependence of α and V_s on $1/p$ for conductor and earthed plate system (50 Hz) (Air at STP)

a   electrostatic field study ($\alpha = \eta.g$)
b   high voltage study
    (i)   R = 4.5 in, r_0 = 1.0 in
    (ii)   R = 5.5 in, r_0 = 1.0 in
    —— predicted $V_s = E_s.\eta.g$. where $E = 27.2 + 13.35/\sqrt{r}$
  o   experimental

In the past, the only major limitation to the application of this empirical approach to switchgear insulation design evaluation has been the lack of reliable breakdown information. Ryan et al have made significant contributions (6.11, 6.20) in this area and have shown that critical breakdown gradients (E_{50}) and highest withstand gradients (E_w) in air and SF_6 can be accurately expressed by relationships of the form

$$rE = K_1 p r + K_2 \qquad \ldots \qquad \ldots \qquad (6.2)$$

where K_1 and K_2 are constants and r is the radius of the inner cylinder. The real virtue of this simple empirical breakdown estimation technique is the fact that one can often consider practical electrode systems to be perturbations of some simple geometry - e.g. Fig. 6.17 is a perturbed coaxial cylinder system. Predicted and experimental sparking voltages have been compared for numerous electrode systems using perturbation field techniques and excellent agreement has been demonstrated (e.g. see Table 6.3).

6.5.2.2 Semi-empirical approach

As an alternative to the above empirical method, the author and his colleagues have investigated a so-called semi-empirical breakdown estimation method, based on streamer theory (6.4). For air, the criterion for minimum breakdown voltage was given by Pederson et al (6.25, 6.26) as

$$\ln \alpha_x + \int_0^x \alpha \, dx = \ln\alpha + \alpha x \qquad \ldots \qquad (6.3)$$

where α_x is the numerical value of α, Townsend's first ionisation coefficient, at the head of the avalanche, of length x, at which the critical ion number is reached in an electron avalanche, in a non-uniform field, streamers are formed resulting in corona or breakdown.

The left-hand side of equation 6.3 is evaluated from the electrostatic field distribution obtained using the general field analysis program for practical arrangements or precise equations for standard geometries (6.19 - 6.22). The right hand side of equation 6.3 is computed from existing empirical breakdown potential gradient data for uniform fields of gap x. The values of α used in both sides of equation 6.3, for the examples discussed in previous studies (6.19, 6.20) have been published by earlier workers (6.4, 6.20).

General purpose digital computer programs have been developed which are capable of solving equations of the form of 6.3 to predict, with acceptable accuracy, the minimum breakdown voltage, V, for a wide range of standard and practical arrangements in air, N_2 and SF_6 for pressures in the range 1 - 5 atmospheres (6.19, 6.20 and 6.22). An example of these techniques is given in Table 6.3 which compares experimental and estimated breakdown voltages in air.

Table 6.3   Breakdown data for hemispherically ended rod/plate arrangement (r = 12.7 mm)   (Ref. 6.20)

	Gap g mm	Breakdown voltage kV at relative density,		
		3	5	7
Experimental		100	158	210
Empirical*	20	102	158	214
Semi-Empirical		103	162	220
Experimental		127	200	259
Empirical*	40	126	197	267
Semi-Empirical		129	203	276
Experimental		144	227	288
Empirical*	80	143	222	301
Semi-Empirical		142	227	310

* Estimate using equation 6.1, together with equation of form r.E. = K_1p.r. + K_2 derived from concentric sphere hemisphere data.

6.5.2.3 Summary

A brief indication has been given in this section of numerical field techniques which have now found widespread application in GIS and switchgear design. The simple break-down estimation methods (empirical and semi-empirical) have now been developed to such a degree that minimum breakdown voltages of practical GIS design layouts can be estimated at the design stage without recourse to expensive development testing. Such derived voltages are minimal withstand levels attainable under practical conditions.

Acknowledgements

The author wishes to thank the Directors of NEI Reyrolle Ltd for permission to publish this document. He also acknowledges the generous assistance given by his colleagues.

REFERENCES

6.1 Legg, D., 1970, 'High-voltage testing techniques', Internal Reyrolle Research Report.

6.2 Diesendorf, W., 1974, 'Insulation co-ordination in high voltage electric power systems', Butterworths.

6.3 IEEE Committee report, 1974, 'Sparkover characteristics of high voltage protective gaps', IEEE Trans. Power Apparatus and Systems, 93, 196-205.

6.4 Craggs, J.D., and Meek, J.M., 1978, 'Electrical breakdown of gases', J. Wiley & Sons.

6.5 Malik, N.H., and Qureshi, A.H., 1979, 'A review of electrical breakdown in mixtures of SF_6 and other gases', IEEE Trans. Electr. Insul., Vol EI-14, 1-14, February.

6.6 Flurscheim, C.H., 1982, 'Power circuit-breaker theory and design', Peter Peregrinus Ltd.

6.7 Ragaller, K., 1978, 'Current interruption in high voltage networks', Plenum Press.

6.8 Ryan, H.M., and Milne, D., 1983, 'Dielectric testing of GIS: Review of test procedures and evaluation of test results', Paper 33-83 (SC) 05-4, CIGRE Colloquium, Edinburgh, SC33, June 6-8th.

6.9 Mosch, W., and Hauschild, W., 1979, 'High voltage insulation with Sulphur Hexafluoride', VEB Verlag Technik Berlin.

6.10 Boggs, S.A., Chu, F.Y., Hick, M.A., Rishworth, A.B., Trolliet, B., and Vigreux, J., 1982, 'Prospects of improving the reliability and maintainability of EHV gas insulated substations', CIGRE, 23-10.

6.11 Ryan, H.M., and Watson, W.L., 1978, 'Impulse breakdown characteristics in SF_6 for non-uniform field gaps, CIGRE Paper 15-01.

6.12 Laghari, J.R., 1981, 'Review - A review of particle contaminated gas breakdown', IEEE Trans. on Electrical Insulation, Vol. Ei-16, 388.

6.13 Ishikawa, M., and Hattori, T., 1982, 'Voltage-time characteristics of particle initiated breakdown in SF_6 gas', 3rd Int. Symp. on Gaseous Dielectrics. Knoxville, USA, Paper No. 28.

6.14 Eteiba, M. G., and Rizk, A.M., 1983, 'Voltage-time characteristics of particle-initiated impulse breakdown in SF_6 and SF_6-N_2', IEE Trans. on Power Apparatus and Systems, Vol PAS-102, 5, 1352-1360.

6.15 Galloway, R.H., Ryan, H.M., and Scott, M.F. 1967, 'Calculation of electric fields by digital computer', Proc. IEE, 114 (6), 824-829.

6.16 Ryan, H.M., and Walley, C.A., 1967, 'Field auxiliary factors for simple electrode geometries', Proc. IEE 114, (10), 1529-1534.

6.17 Mattingley, J.M., and Ryan, H.M., 1971, 'Potential and potential-gradient for standard and practical electrode systems', Proc. IEE, 118, (5), 720-732.

6.18 Ryan, H.M., 1973, 'Prediction of electric fields and breakdown voltage levels for practical 3-dimensional field problems', Eighth Universities Power Engineering Conference, University of Bath.

6.19 Blackett, J., Mattingley, J.M. and Ryan, H.M., 1970, 'Breakdown voltage estimation in gases using semi-empirical concept', IEE Conf. Publ. 70, 293-297.

6.20 Mattingley, J.M., and Ryan, H.M., 1971, 'Breakdown voltage estimation in air and nitrogen', NRC Conference on Electrical and Dielectric Phenomena, Williamsburgh, USA, November.

6.21 Ryan, H.M., 1967, 'Prediction of alternating sparking voltages for a few simple electrode systems by means of a general discharge-law concept', Proc. IEE, 114, (11), 1815-1821.

6.22 Scott, M.F., Mattingley, J.M., and Ryan, H.M., 1974, 'Computation of electric fields: Recent developments and practical applications', IEEE Transactions on Electrical Insulation, Vol. EI-9, No. 1, March, 18-25.

6.23 Binns, D.F., and Randall, T.J., 1967, 'Calculation of potential gradients for a dielectric slab placed between a sphere and a plane', Proc.IEE, 114, (10), 1521-28.

6.24 Storey, J.T., and Billings, M.J., 1969, Determination of the 3-dimensional electrostatic field of a curved bushing', Proc. IEE, 116 (4), 639-643.

6.25 Pederson, A., 1967, 'Calculation of spark breakdown or corona starting voltages in non-uniform fields', IEEE Trans. Power Apparatus and Systems, 86, 200-206.

6.26 Pederson, A., 1967, 'Analysis of spark breakdown characteristics for sphere gaps', ibid, 975-978.

Chapter 7

Properties and applications of liquid insulants

F. W. Waddington

7.1 INTRODUCTION

Petroleum oils, esters, chlorinated liquids, silicones, synthetic hydrocarbons and fluorinated hydrocarbons are all used in everyday electrical equipment, such as transformers, capacitors, cables and switchgear. Liquefied gases, electronegative fluids and refrigerants are used in the more specialised applications of superconducting magnets, generators, non-flammable switchgear and vapour-phase transformers.

Considering each of these fluids in more detail:

7.1.1 Petroleum Oil

Petroleum oil is still the most widely used insulant or dielectric, its most common application is in the form of a very thin oil, almost a kerosene in its characteristic odour. Petroleum transformer oil to BS148 (7.1) is the prime insulant in most high voltage transformers, switchgear, bushings and cables and its production and maintenance in the U.K. involves the handling of millions of gallons per annum of this oil. As a fraction from crude petroleum in-between fuel oils and lubricants it is a cheap commodity still in plentiful supply. Other uses of this thin oil in the textile industry e.g. as a light spindle lubricant has been greatly reduced in modern times. Heavier grades of petroleum find limited applications in certain transformers where efficient cooling is not required and a moderate degree of fire resistance has to be achieved. A very viscous almost solid impregnant in the form of penetrol oil, or petroleum jelly, or even bitumen, is used in impregnation of cables, terminal boxes etc., to prevent ''draining'' from high levels to low levels during service.

7.1.2 Esters and Synthetic Hydrocarbons

Esters and synthetic hydrocarbons (7.2) are man-made improvements on petroleum oil and can be chemically tailored to produce more precise properties such as a high thermal stability or fire resistance when used in transformers, switchgear etc. and a high resistance to discharge or impulse breakdown when used in capacitors or cables. In these respects they can be superior dielectrics to petroleum oil but are considerably more costly being 3 to 4 times the price of petroleum. Polybutenes (7.3) and alkyl benzenes also have certain advantages in cables and capacitors.

7.1.3 Silicones and Fluorinated Oils

Silicones and fluorinated oils can further enhance the thermal stability properties as dielectric fluids but have certain disadvantages. In the case of silicone this fluid does produce considerably improved fire resistance but can suffer from poor heat transfer and a semi-conducting property on discharge (7.4) whereas fluorocarbons, though excellent dielectrics and having excellent fire resistance properties, are at present very expensive at 40 to 50 times the price of petroleum. Also, on discharge the fluorinated gases produced have still to be thoroughly identified and toxicologically classified (7.5).

7.1.4 Chlorinated Liquids and Phosphate Fluids

Chlorinated liquids include the now notorious askarels or chlorinated diphenyls were relatively cheap (twice the cost of petroleum), very fire resistant, thermally stable and discharge resistant dielectrics but have recently been found (7.6) to be highly toxic to certain lower forms of life and to be severely bioaccumulating in the earth's environment. Phosphate esters (7.7) are also relatively cheap and their initial high toxicity has now been largely resolved. They are, however, poor dielectrics and their environment hazards are still questionable.

7.1.5 Electronegative Fluids or Refrigerants

Electronegative fluids or refrigerants can also be wholly or partly chlorinated and may suffer from the above toxicity. A great deal of work is in hand (7.8) at the present time to investigate this property and to examine the nature of gases (particularly hydrochloric acid vapour) which can be evolved on discharge. The advantages of such fluids as the refrigerant Arcton 12 (tri-fluorotrichloroethane) lies in their use as vapourisation cooling dielectric in transformer cooling or arc quenching; also the electronegative (electron absorbing) properties of the fluorine or chlorine molecule as in the gaseous compound $SF6$ (sulphur hexafluoride) make these fluids excellent arc/ discharge suppressors.

7.1.6 Liquefied Elemental Gases

Liquefied elemental gases have very specialised applications as dielectrics to function at very low temperatures such as superconducting magnets in high energy particle machines or in a future generation of highly efficient rotating machines. Liquid helium (7.9) is the

only dielectric liquid cool enough (boiling point 4.2K) to
allow efficient use of superconductors at present in use.
Liquid hydrogen (boiling point 14-20K) can be used to
utilise the high conductivity of aluminium at 20K and liquid
nitrogen (boiling point 77K) is used as an intermediate
cooler. All are good dielectrics at these temperatures
having high break-down strengths (40-100kV/mm) and low
permittivities of 1 to 1.5.

7.2 MANUFACTURING METHODS AND NATURAL SOURCES OF LIQUIDS

7.2.1 Mineral Oils

Still the most widely used of dielectric fluids
petroleum oils can be classified according to the proportion of three types of naturally occurring chemical
structures originating from the crude oil used in their
manufacture (7.10). Paraffinic oils are those mainly
containing straight and branched chain hydrocarbons;
naphthenic oils contain large numbers of cycloparaffins and
aromatic oils contain ring molecular structures with roughly
equal numbers of carbon and hydrogen as in benzene (e.g.
C_6H_6). The first type of natural petroleum oil is the most
resistant to oxidation and have a very good response to
antioxidants used to improve this property. On the other
hand the aromatic oils containing many "benzene" type
structures are manufactured as resistant to discharge. The
former paraffinic oils are preferred for applications such
as switchgear or where low voltage stress and high temperatures are involved whereas the "aromatic" oils are
preferred for high voltage applications in cables, or high
voltage transformers.
There has been no major change in the method of refining and distilling mineral oil for several decades though
the use of acid refining has been abandoned for solvent/hydro
refining to remove impurities and wax from the crude
petroleum.

7.2.2 Synthetic Oils

It appears an obvious improvement over petroleum oil
to attempt to synthesise the precise molecular structure
required to produce the dielectric fluid for any application. At the same time certain properties not available in
petroleum oil such as low-flammability could be built into
the structure of the fluid.
The manufacture of esters (7.2) and synthetic hydrocarbons are good examples of this molecular structuring
of a fluid.

7.2.2.1 Synthetic Esters and Hydrocarbon Manufacture.
Esters are synthetic fluids which are oily liquids and can
be made with exact proportion of paraffinic or aromatic
characteristics and can also be made more thermally stable
and with a high degree of low flammability (7.2). The
term 'esterification' refers to the combination (often at

very high temperatures up to 300°C) of an alcohol with a slightly acidic organic acid. Glycerine and coconut oil are two everyday examples or alcohols and acids, but electrical grade esters are manufactured from more pure chemicals such as pentaerythritol and heptanoic acid, or butanol and a phthalic acid. The result of such esterifications are white oils which if symmetrical in their molecular structure and free of ionic contamination can have excellent electrical characteristics (7.2).

Hydrocarbons and oxyhydrocarbons can be manufactured (7.11) from ethylene and other olefine gases which can be polymerised to liquids under very high pressure and/or temperature. Ethers are another class of oxyhydrocarbons which can be made by reacting quite common compounds such as phenol with bromobenzene, analine and chlorphenols. The polyphenylethers which result from this latter type of reaction are some of the most heat resistant and stable liquids known at the present (7.12).

7.2.2.2 Silicones, phosphate esters, fluorocompounds and chlorinated biphenyls.

These are all complex synthetics which incorporate a "foreign" atom into the conventional "organic" carbon/hydrogen/oxygen structure to produce an improved or modified property. The consequence of this can be a remarkably improved property (7.4) such as thermal stability but problems of toxicity, environment acceptance or an electrical weakness can develop.

Manufacturing methods are usually energy intensive requiring the electrolytic production of fluorine, chlorine, silicon fluoride or phosphoric acid as a starting point and the synthesis can follow several stages of purification or fractionation.

It is particularly important with the manufacture of the synthetics that ionic impurities are thoroughly removed and that moisture is avoided. The risk of hydrolysis of the halogenated /phosphate compounds to strong acids on electrical discharge or overheating often necessitates that an acid scavenger such as epoxide or metal complex (e.g. tetraphenyltin) (7.4) is incorporated into the fluid to absorb traces of acid products.

7.2.2.3 Refrigerants and Elemental Liquefied Gases.

The manufacture of refrigerants such as Arcton or Freon again require the reaction of elemental fluorine or chlorine with sulphur, hydrocarbon gases or oils. This is again an energy intensive process requiring careful purification if the resultant gas or liquids are to be used as dielectrics.

Elemental gases (7.13) are formed by simple fractionation of air, as a residue from combustion of air (i.e. Nitrogen), as a chemical reaction product (e.g. H_2 from acid and metal) or separated from natural gas (e.g. Helium).

7.3 PROPERTIES REQUIRED FOR POWER EQUIPMENT

In considering the properties of dielectric fluids we must understand the function required in any particular type of electrical component.

7.3.1 Transformers

Transformers require a fluid to insulate often at very
high voltages, the fluid is also required to act as an
efficient coolant or heat transfer media. To maintain
energy efficiency the fluid should have a low loss tangent
and a high resistivity and in very high voltage applications
should be resistant to partial discharge. In certain cases,
particularly for indoor application the fluid should be of
low or non-flammability and because of the large volumes
often employed, environment and health authorities require
that the fluid is non toxic and biodegradable. Finally
the fluid is preferred to have good arc quenching properties
good lubrication, a low gas absorption and a low coefficient
of expansion.
The two fluids which most closely meet the majority of
these conflicting requirements are still petroleum oil to
BS148 and more recently the transformer esters to IEC
10B/30 (7.14). Other synthetic fluids can have weaknesses,
electrically, thermally or in respect of toxic/biodegrad-
ability properties.

7.3.2 Capacitors

Exceptional resistance to discharge, high inception
voltages and gas absorbing properties are of paramount
importance in a capacitor (7.15). Low dissipation factor
and low viscosity are also of importance. In certain cases
a low flammability is becoming of greater importance and
low toxicity is now considered to be a desirable property
Here, the aromatic benzene ring structure has an
inherent ability to absorb partial discharge energy and
free electrons. This follows from the resonant electron
sharing nature of the unsaturated C H ring structure which
is considered an essential component of a capacitor fluid.
A diphenyl or alkyl diphenyl with two benzene rings in each
molecule is the preferred fluid for discharge but if low
flammability and low toxicity is required a single benzene
ring in the form of a symmetrical ester structure is a
promising compromise. (7.2)

7.3.3 Cables

Cables have a similar requirement to capacitors, here
dodecylbenzene (7.16) has been used for many years being a
low viscosity highly aromatic cable fluid. DDB is, however,
of high flammability and particularly for indoor application
the use of the less flammable symmetrical aromatic esters
are now being considered. For large underground power
cables petroleum oil to BS148 is still the most commonly
used dielectric.

7.3.4 Switchgear

Arc suppression in a fluid system is of major importance, but low flammability is now being more actively sought after. Esters are a compromise fluid superior to petroleum oil in flammability. Completely non-flammable fluids are available but these are invariably halogenated and products of arcing can be corrosive or toxic; more work is required in this area.

7.4   DIELECTRIC AND ELECTRICAL PROPERTIES OF FLUIDS

7.4.1   Breakdown Strength

Liquids can vary widely in electrical breakdown strength; power frequency breakdown (usually 50 or 60 c/s to IEC ASTM D1816 or BS 148) is measured across a 2.5 mm gap between spheres or hemispheres. Values reached depend critically on contaminants in the fluid such as fibres, dust particles or moisture particles which can align to form bridges; values can vary from less than 10kV to as high as 100kV with a very dry highly filtered fluid (Fig. 7.1). Certain fluids can generate conducting or semi-conducting products on arcing or discharge, these can also lead to very low repeat breakdown by bridging. Such fluids should not be used in systems where discharge or arcing is likely to occur (7.17).

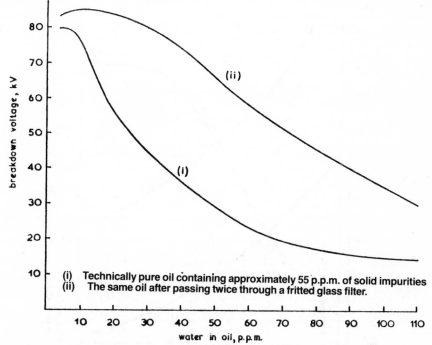

(i)   Technically pure oil containing approximately 55 p.p.m. of solid impurities
(ii) The same oil after passing twice through a fritted glass filter.

Fig.7.1   Breakdown voltage of a transformer oil at 25°C between spheres 12.5mm diameter, 3.5mm apart (Reference 7.20)

Switching surge and impulse breakdown are believed
(7.18) to depend more on the ability of the fluid system
to rapidly grow gas bubbles at the site where electrons
are injected into it. Increase in fluid viscosity, (7.19)
lowering of gas content, introduction of fibre matrix or
electronegative atoms are all ways of improving impulse
strength.

7.4.2 Dielectric loss and volume resistivity

At low frequencies the power loss in a mobile liquid
(viscosity 1 to 100 c/s at 20°C) is largely controlled by
the conductivity (G) part of the loss equation i.e.
according to the function:-

$$Tan\,\delta \simeq \frac{G}{\omega C}$$

where ω is angular frequency ..... (7.1)
and C is capacitance

Dielectric loss by dipole contribution at frequencies
of 50 or 60 c/s only appears important in more rigid
systems with large dipoles such as waxes or very viscous
oils. Both loss and resistivity in a liquid are therefore
often found to be largely controlled by conducting
particles, moisture and other ionic contamination. The
removal of these contaminants by molecular filtration for
example can produce a dramatic improvement in loss and
resistivity values (Fig. 7.2). (7.4).

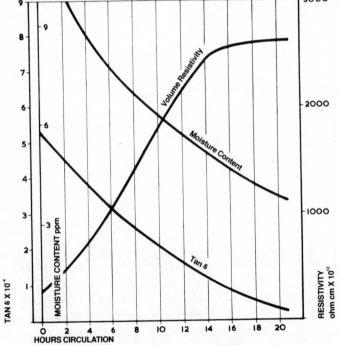

Fig.7.2   Improvement in BS148 Transformer Oil by
circulation through a Molecular Sieve Filter

7.4.3 Discharge Resistance

The basic circuit for discharge detection is shown in
Fig. 7.3 reproduced from Sillars (7.20) 1973. The discharge
developing in the insulation under test as the voltage is
increased is displayed on the c.r.o. The voltage at which
discharge first appears, the inception voltage Vi, is an
important parameter and in complete systems such as a
bushing or capacitor the voltage at which it ceases V_e (ie
the extinction voltage) is also important.

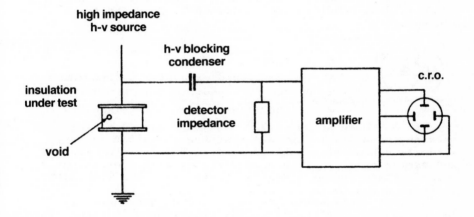

Fig.7.3   A simple circuit for discharge detection using
            a Cathode Ray Oscillograph (c.r.o.)

Discharge energy E in joules can be simply related to
a discharge level Q in picocoulombs at any peak voltage
Vi by the function:-

$$E \;\; = \;\; \tfrac{1}{2} \;\; Q.V_i \;\; pk \; \ldots \ldots \; 7.2$$

This in turn can be related to the breakdown of the
liquid to form gaseous products (7.21). It can be shown
that the quantity of gas formed depends mainly on the energy
involved and less on the type of fluid. The major gas
produced from the fluids, e.g. oils, esters and synthetic
hydrocarbons, used in highly stressed systems is hydrogen
The formation of gaseous bubbles of hydrogen can in turn
initiate discharge.
        V_i Levels can be increased and discharge energy de-
creased by the use of strongly gas (hydrogen) absorbing

fluids, as described in section 7.3 such fluids at present depend on a high concentration of benzene ring structures.

The efficiency of a fluid to absorb discharge gases can be measured using a Pirelli test equipment; this consists of an arrangement whereby a corona or partial discharge from a 10 - 20kV, 50cs source is passed between a centre electrode over the surface of a fluid to the wall of a sealed glass tube. Gases evolved from or absorbed by this surface of the liquid can be measured by pressure changes in the gas space above the liquid. Details of Pirelli testing are given in Chapter 11.

7.5   THERMAL TRANSFER CHARACTERISTICS

In a liquid filled system such as a transformer or cable, heat is transferred mainly by convection. Convection under natural conditions in a liquid is given by the following function:-

Where

K   is thermal conductivity
a   is the liquid expansion coefficient
c   represents specific heat per unit volume
v   is the liquids kinematic viscosity

It has been shown (7.22) that the power n varies between $\frac{1}{3}$ and $\frac{1}{4}$. It follows that heat transfer is very strongly dependent on the thermal conductivity of the liquid and varies to a lesser degree directly with coefficient of expansion and specific heat and inversely with viscosity. There is not a large difference in specific heat or coefficient of expansion between liquids but there is a very large variation in the viscosity of a liquid. It therefore follows that heat transfer in a static liquid (i.e. natural cooling) is mainly controlled by its thermal conductivity and its viscosity and to a lesser degree by expansion coefficient a.

In a system where the liquid is pumped or otherwise force-circulated around the components this is even more the case in that expansion coefficient which was involved in the above function as a liquid's own internal device for enhancing liquid movement no longer plays an effective part in a pumped system and can be ignored.

The choice of a liquid dielectric which also has to function as a heat transfer mechanism is particularly important in certain designs of transformers, reactors, rectifiers or cables where large amounts of electrical energy is being transmitted. Examples of these occur in Railway Locomotive transformers, furnace transformers and high power cables. Design data given in Figs. 7.4, 7.5 and 7.6 (7.23) is therefore critical to the electrical engineer.

Fig. 7.4 shows the variation in thermal conductivity of various fluids; it is important to note that K varies considerably with temperature in certain cases increasing in others decreasing. Obviously an increasing value is preferable for systems liable to operate continuously at

high temperatures; on the other hand problems with heat
transfer can occur at very low temperatures (i.e. when other
parameters such as very high viscosity aggravates heat
transfer) and poor thermal conductivity could lead to
localised overheating or even an electrical "burn-out".

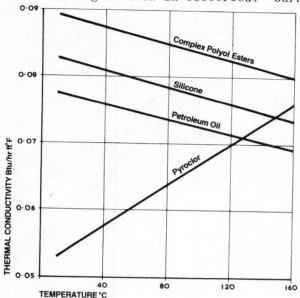

Fig.7.4   Thermal Conductivities of Fluids

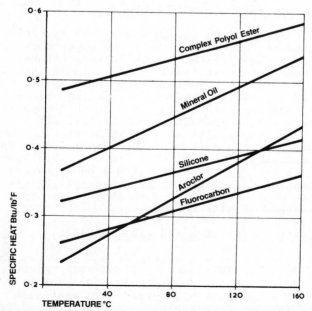

Fig.7.5   Heat capacity of Fluids

Fig. 7.5 illustrates typical variation in specific heats c of different dielectric fluids. Though c only effects heat transfer to the first power, it can be seen that there can be considerable variations in this value with temperature. Again the argument presented for conductivity will also apply to specific heat. It is, however, important to note that specific heat is a value quoted per unit weight (e.g. per gm) so that fluids with a high density such as the fluorocarbons or PCBs have the added advantage of giving more weight per unit size and therefore proportionately more heat transfer.

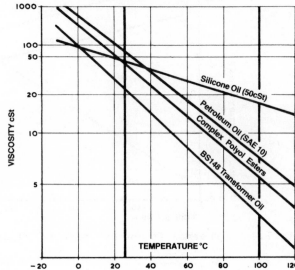

Fig.7.6
Viscosity of Fluids

Viscosity (Fig. 7.6) though again only appearing in the heat transfer function to the first power does vary enormously in a liquid with changes in temperature. It is one of the most important features of a liquid that as the temperature increases there is a very rapid fall off in viscosity; this greatly enhances the cooling effect of a fluid which together with an increase in heat capacity of the fluid more than compensates for the fall-off in thermal conductivity of many fluids at high temperature. One particular fluid which does not exhibit this rapid deteriration of viscosity are the silicone fluids. This viscostatic property shown in Fig.7.6coupled with a poor heat capacity (specific heat) and falling thermal conductivity can result in severe overheating problems with this fluid especially in pumped systems at high temperatures.

7.6 RESISTANCE TO IGNITION AND FIRE PROPAGATION

Certain electrical equipment for use in areas where a fire or explosion would cause a serious risk to nearby personnel, or inflammable plant/chemicals, must be designed to present the least possible hazard. To meet such requirements various Fire/Explosion authorities (7.24)(7.25) have devised certain safety specifications and standards.

For example - The National Fire Protection Association of the U.S.A. (7.24) issue a comprehensive specification for all electrical equipment and materials. Of particular interest the specification for a low flammability dielectric fluid for indoor transformers, (NFPA 450-23-1978) states that the fluid should have a fire point (BS 4689, IP30/63) greater than 300° and the fluid should be non-propagating. This later property is generally tested by a spray ignition test (either Factory Mutual USA Approval Standard, Class No. 6930 or the UK Health and Safety Executive test No.570/1970) where a high pressure jet of the liquid is ignited by a blow torch and must self extinguish after removing the source of ignition in a specified number of seconds depending on the pressure and spray temperature used (usually 1000 p.s.i. and 140°F).

The Factory Mutual Corporation (7.25) have more recently issued a "heat release" test (FM5-45/14-85, 1979) where the convective and radiative heat release values are measured during the combustion of large volumes of the fluid (e.g. 45 gallons). The values found determine the safe distance a fluid filled electrical appliance may be safely situated from combustible objects or personnel. Very few fluids will not ignite and those that are genuinely non-inflammable toxic fumes may still be evolved on exposure to heat and the fluid may have toxic properties. Other methods of clarifying the Fire Hazard from a fluid include the measurement of specific properties including heat of combustion, specific heat, thermal conductivity flash point and autoignition values.

A fluid with a high specific heat and high thermal conductivity will be difficult to ignite, i.e. to produce an input of heat at sufficient rate to raise its temperature to above its fire point.

Methods used to test fluids for flash point and fire point are described in a later chapter, but in addition to these and the more practical full-scale spray and fire tests described above Authorities are continually seeking for a laboratory test to assess the fire hazard of a fluid. Of those mentioned earlier the oxygen index value has been studied in considerable detail at the U.K. ERA Institute of Technology (7.26). A line drawing of the combustion cell is given in Fig. 7.7. Where a central glass wick terminating at fluid level in the cell serves to help ignite and maintain a "flame" of the fluid under test; cooling fins on the cell help maintain a constant fluid temperature. Surrounding the unit with a glass envelope into which nitrogen oxygen mixture with a progressively reduced oxygen content can be introduced enables a measurement to be made of the level when combustion ceases and the "flame" is extinguished. Oxygen index values are calculated as that percentage of oxygen present in nitrogen at which combustion just continues. This varies from 18% to 20% for a good low flammable fluid such as silicone and certain esters to less than 15% for highly inflammable petroleum products. The

considerable difference in flame size between a polyolester
fluid and a petroleum oil to BS148 is illustrated in Fig.
7.8.   This illustrates normal combustion in air (20% oxygen)
prior to surrounding the flame cell with the glass envelope
and reducing the oxygen content to extinction values.

   This work is in addition to conventional testing for
flash point and fire point which is described later in
Chapter 11.   The fire resistance of a fluid is, however,
the combination of all its relevant properties including
the temperature it ignites, the speed of ignition and the
heat evolved after ignition.   All factors must be taken into
account in assessing the hazard associated with a particular
fluid.

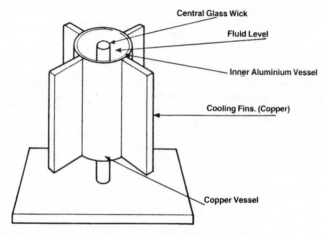

Fig.7.7   ERA Test Cell for Oxygen Index Test (7.26)

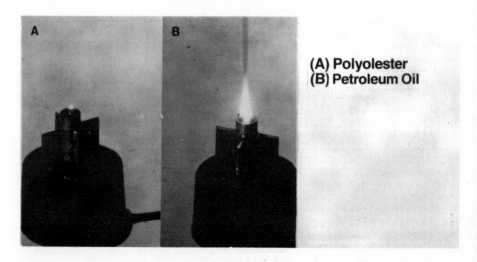

Fig.7.8   Comparison of Flame Sizes

TABLE 1 Gas and solid residue yields from arcing
in various fluids

FLUID	Gas yield ml/joule	Solids yield gm/joule
Petroleum Oil	38	0.023
Silicone	30	0.065
Complex Polyol Esters	10	0.002
Pyroclor	30	0.131
Fluorocarbons	6	0.001

7.7 ARC SUPPRESSION PROPERTIES OF LIQUIDS

The ability to suppress a power arc is a critical property when a liquid is used in a circuit breaker, tapchanger or switchgear; and is a safety requirement in transformers, cables and capacitors where a power arcing fault could accidentally occur. There are two basic mechanisms which can explain the surprising ability of even highly inflammable petroleum oils to extinguish high power arcs drawn between metal electrodes under the surface of the liquid. In contrast arcing above or along the surface will be a major disaster with an inflammable fluid.

7.7.1 Arc Suppression in Hydrocarbons and Esters

Hydrocarbons such as a petroleum transformer oil to BS148 or esters such as a pentaerythritol transformer ester (IEC 10B30) are excellent arc suppressors. The mechanism has been studied by many authorities (7.27) (7.28) who generally agree that the initial arcing reaches temperatures of over $3000^{\circ}C$ at which temperature the oil molecules are decomposed to give a plasma of hydrogen, carbon and oxygen atoms and the arc is maintained in a bubble of these gases. These atomic gases will reform in the cooler parts of the plasma to give mainly hydrogen gas (H_2) with lesser quantities of methane (CH_4) acetylene (C_2H_2) and small quantities of CO and CO_2. It is the cooling effect of these endothermic (heat absorbing) reactions and the high thermal conductivity of the large quantities of hydrogen formed which is believed to cool and control the arc. In a switch or circuit breaker the contacts are in the meantime separating at a high velocity. To successfully extinguish an arc all that is required in an AC power system is that this gas plasma is cooled sufficiently whilst the arc passes through a voltage zero to establish an insulating gas bubble rather than a conducting plasma so that the arc does not restrike during a further half cycle at a voltage maximum. Obviously the more rapidly this occurs (i.e. the minimum number of half cycles) the less energy is dissipated.

Oils are also chosen for switch applications which produce minimum quantities of black carbon residues on arcing. This has been found to be simply proportional to the carbon content of the fluid. Paraffinic oils contain less carbon i.e. $(CH_2)^n$ than aromatic $(CH)^n$ and esters $(CH_2O)^n$ contain the least carbon per unit mass. Esters are now finding a wider application as a switch fluid for this reason.

7.7.2 Arc Suppressions by Electronegative Fluids

Electronegative fluids are generally so called because they contain an atomic structure which can accept further electrons (or other charge particles.) Sulphur hexafluoride both in the gaseous and liquid form is a classic example of this; other compounds such as carbon fluorides and carbon chlorides also exhibit various degrees of electronegative character.
Under conditions of arcing these fluids do not decompose to give hydrogen (which is not present) and appear to rely on heats of vaporisation and simple production of lower molecular weight fluoro or chloro compounds to cool the arc. (7.29) The establishing of a non-conducting gaseous plasma is, however, aided in this case by the electron scavenging characteristic of these fluids.
Electronegative fluids can be formulated to contain less carbon i.e. CCL contains only one carbon atom in a large heavy molecule compared to hydrocarbons or esters. They also have the advantage of complete lack of flammability.
Fluorine and halogen compound do suffer from a disadvantage that toxic or corrosive products can be formed on arcing and the fluids must always be used in a sealed system.
Table 1 shows (7.23) the relative carbon or solid residues and gas yields from a range of modern dielectric fluids at present being used in switch systems. The lower the solid or gas yield the more compact the unit which could be designed.

REFERENCES

7.1   BS148 (1972) Specification for insulating
        oils for transformers and switchgear.
        Publication No. Gr. 7.

7.2   Waddington, F.B. (1979) "New Synthetic
        Ester Fluids." pp 8-11 IEEE Publication
        No. CH1510-7 EIC Conference, Boston, October

7.3   Wilson, A.C.M. (1980) "Insulating liquids
        their use, manufacture and properties."
        Peter Peregrinus Ltd., Stevenage, U.K. and
        New York.

7.4   Birks, J.B. (1960) "Modern dielectric
        Materials." Heywood & Company Ltd., London.

7.5   Goodman, D.M. and Waddington F.B. (1982)
         Seventh International Conference on Gas
         Discharges and Application. IEE London.

7.6   Clark. F.M., Coursey, P.R., Liebscher, F.,
         Potthoff, K.W. and Viale, F., (1958).
         "Nonflammable liquid impregnants"
         CIGRE 4-14 June.

7.7   Hatton, R.E. (1962) "Introduction to hydraulic
         fluids." Chapman & Hall, London.

7.8   Coleman, C.R. (1965) Proceedings IEE 112(3) March.

7.9   Meats. R.J. (1972) Proceedings IEE 119 pp760-6:
         73,231.

7.10   Wilson, A.C.M. (1980) Petroleum mineral oils
         -Chapter 3. "Insulating Liquids" Peter Peregrinus,
         Stevenage, U.K. and New York.

7.11   Renfrew, A. and Morgan. P(1959) "Ethylene
         polymers." Iliffe & Sons Ltd. London.

7.12   Mahoney, C.L. and Barnum,E.R. (1962)
         Polyphenyl ethers. "Synthetic lubricants."
         Sanderson, R.G. and Hart, A.W. New York.

7.13   Furnas, C.C. (1943) "Manual of Industrial
         Chemistry" D.Van Nastrand Co. Inc. New York

7.14   IEC publication 10B/30 (1982) Central Office,
         IEC - Geneva, Switzerland.

7.15   Parkman, W. and Dickson, M.R. (1978) "Impregnants
         for power capacitors." BEAMA International
         Conference Electrical Insulation. Brighton
         2 -5 May.

7.16   Reynolds, E.H. and Black, R.M. (1972). Proc. IEE
         119(4) pp497-504.

7.17   Dowling, D.J. (1960) "Modern Dielectric
         Materials" p. 147-148. Heywood and Co. London.

7.18   Singh, B., Chadband, W., Smith, C.W. and
         Calderwood, J.H. (1982) J. Physics D. 5, pp
         1457-64, 74.

7.19   Krasuki, Z. (1962) Proc. IEE 109E Suppl.22pp
         435-9, 74.

7.20   Sillars, R.W. (1973) "Electrical Insulating
         materials and their application." Peter
         Peregrinus Ltd. Stevenage.

7.21   Waddington, F.B. (1977) "Fault detection in
        power engineering." GEC Journal of Science
        and Technology. May

7.22   Stoever, H.J. (1941) "Applied heat transmission,"
        35 (McGray-Hill).

7.23   Waddington, F.B. (1978) "Recent developments
        in the field of liquid dielectrics." BEAMA
        International Electrical Insulation Conference. May

7.24   National Fire Protection Association (1978)
        2-5 May, Brighton. "Code of Practice."
        Section 450-23 pp. U.S.A.

7.25   Factory Mutual Association (1979)
        "Specification for less flammable
        transformer fluids." FM545/ 14-85

7.26   Day, A.G. (1974) "The versatility of the
        oxygen index test." IEE Symposium 6
        February 1973, Session 1, Paper 3

7.27   Reece. M.P. (1977) Chapter 2 "Physics of
        circuit breaker arcs." C.H. Flurscheim
        "Power circuit breaker theory and design"
        Peter Peregrinus Ltd., Stevenage.

7.28   Von Engel (1955) "Ionised gases." Clarendon
        Press, Oxford.

7.29   Maggi. E. (1977) Chapter 7. "SF6 circuit
        breaker theory and design" - Peter Peregrinus
        Ltd., Stevenage.

Chapter 8

Production and applications of solid insulants

J. Heighes

8.1 INTRODUCTION

In attempting to describe the considerable number of solid insulating materials used in electrical power equipment, some method of sub-division is looked for. To consider them by application is not entirely satisfactory, because the same material might be used in a number of different types of equipment. The division between naturally occuring and synthetic materials is satisfactory to a degree but the combination of processed naturally occuring materials with synthetic resins means that this division would only be acceptable with tolerance.

The method of classification which seems most suitable, has resulted from considering how the materials are made. Hence there are sections on material made from the pre-impregnation of the resin support, castings, mouldings, extrusions and pultrusions. Within each section the applications and properties of the material are discussed and an indication is given of how the selection of such materials should be made for specific applications.

8.2 PLASTICS

In order to discuss solid electrical insulating materials, it is necessary to understand the basis of polymer chemistry, upon which a considerable number of insulating materials, is based.

Plastics are usually considered to be materials, produced by synthesis. They may be formed from many identical molecules, for example, the molecules of the gas ethylene, with a boiling point of $-104^{\circ}C$, may be joined together under pressure at high temperature to form polyethylene. This formation of the long chain, called polymerisation, comes about because of the double bonded carbon atom, the polymerisation reaction allowing the carbon atom to maintain its valency of 4 as shown in Fig 8.1. Some polymers still contain double bonds after the polymerisation reaction has taken place and these double bonds may be reacted with some other molecular structure which joins the polymeric chains together. An example of this is unsaturated polyester resin which can be reacted, or cross linked with styrene as shown in Fig. 8.2.

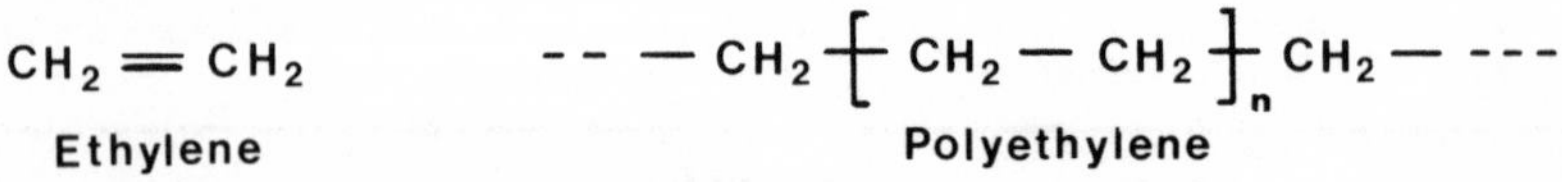

$$CH_2 = CH_2$$

Ethylene

$$-- - CH_2 \left[CH_2 - CH_2 \right]_n CH_2 - ---$$

Polyethylene

Fig.8.1   Example of a thermoplastic

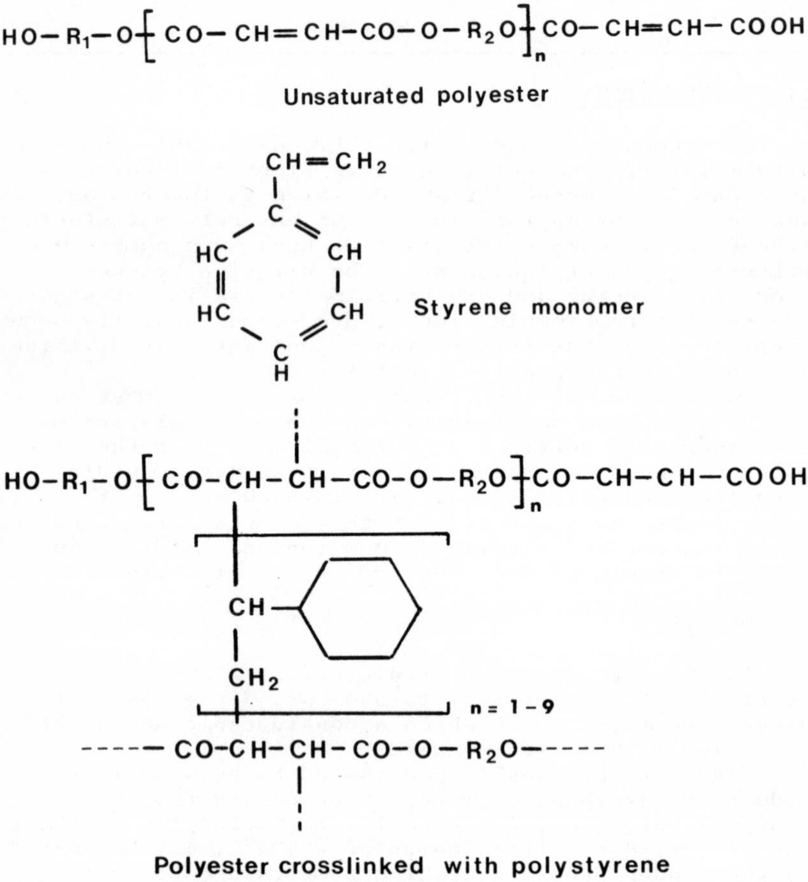

$$HO - R_1 - O \left[CO - CH = CH - CO - O - R_2 O \right]_n CO - CH = CH - COOH$$

Unsaturated polyester

$$CH = CH_2$$
|
$$C$$

HC   CH
‖ |
HC   CH   **Styrene monomer**

C
H

$$HO - R_1 - O \left[CO - CH - CH - CO - O - R_2 O \right]_n CO - CH - CH - COOH$$

CH —

CH$_2$ $n = 1-9$

$$----- CO - CH - CH - CO - O - R_2 O -----$$

Polyester crosslinked with polystyrene

Fig.8.2   Example of a thermoset

These two independant characteristics identify two groups of materials. The first, composed of the linear polymers (not cross linked), will usually soften when heated and is called "thermoplastic". The second group, defined as "thermosetting" is stable to heat and cannot be made to flow or melt. It should be noted however, that such materials will weaken at high temperature and the measurement of the heat distortion temperature is used to assess the thermal capability of the materials.

These two groupings, thermoplastics and thermosetting, are very useful to the insulation engineer although they may not be as acceptable to the polymer chemist (see, for example, Billmeyer, Ref. 8.1).

8.3   LAMINATES AND COMPOSITES

Laminates for electrical insulation, usually consist of layers of a carrier or reinforcement material, impregnated and bonded with a synthetic resin. Some laminates, particularly if reinforced by a glass fibre mat, are produced by placing the reinforcement into a mould, and distributing the resin, usually polyester or epoxide, over the glass fibre. The laminates are then cured under heat and pressure. This method has a number of disadvantages. The only area of laminate which can be produced is the area of the mould, less the edge trimming. Only one laminate can be pressed at a time for each pair of platens of the press; in consequence, the thinner laminates are more expensive than the thicker ones.

These difficulties are overcome when the reinforcement can be pre-impregnated and for this process, there are two requirements. The first is that the reinforcement in its raw state, is mechanically strong enough to be pulled through an impregnating bath and drying chamber. The second is that the resin is capable of being taken to a dry state without being fully cured, a condition known as "B stage".

The impregnated material is usually processed in a continuous length, so that at the end of the pre-impregnated process, it is cut to the required length and built up in stacks to suit the thickness of the laminate. Care must be taken to cool the pre-preg and keep it cool in the stack to avoid the cure becoming too advanced before pressing. Providing precautions are taken, a "pre-preg" can be stored for many weeks before being pressed into a laminate.

The flow characteristic of such a pre-preg is often sufficiently restrained that the laminate does not need to be constrained by a mould. Hence sheet size is only limited by the size of the press and several laminates, if not too thick, may be pressed between each pair of press platens, reducing the cost of pressing.

Typical pressing conditions for laminating, are temperatures in the range of 140°C - 160°C and pressures, in the range of 3.5-7 MN/m^2. A temperature of 160°C may not be high enough for some applications and either a higher pressing temperature or a post cure after removal from the press is necessary.

The reinforcement may be paper, cotton fabric, glass fibre, asbestos, or aromatic polyamide fibre. The glass fibre may be in the form of a woven fabric, a random mat or as uni-directional rovings. The asbestos may be made into a paper, a felt or a fabric.

The resin systems may be phenolic, epoxide, polyester, silicone, melamine and polyimide. Within each of these generic classifications there can be many formulations, designed to produce specific properties, such as flame retardency, arc and tracking resistance often at the expense of other properties.

A considerable range of laminates, is available from many manufacturers, so that it is very difficult to provide a general picture of properties and it is recommended that the individual brochures of manfacturers be consulted. However, Table 8.1 is included in order to indicate the degree to which different reinforcements, the arrangement of the reinforcement and the resin system, may influence the mechanical properties of laminates.

TABLE 8.1 A comparison of the crossbreaking strength (MN/m^2) of various laminates at room temperature

Resin Reinforcement	Phenol Formaldehyde	Epoxide	Unsaturated Polyester
Cellulose Paper	135	140	125
Cotton Fabric	140	170	125
Asbestos Paper	200		
Asbestos Fabric	130		
Glass Fabric	350	450	380
Glass Mat		350	200
Directional Glass Bi-directional	600	750	
Directional Glass Uni-directional	825	1250	
Polyamide		450	

Machined components are often produced from laminates, usually for the purpose of mechanical support or constraint, for example, as slot wedges and spacer blocks in rotating machines, as operating links in switchgear, for printed circuit boards and terminal boards.

Within the heading of laminates, tubes may also be considered, the most extensively used probably being phenolic paper tubes. These are made from a pre-coated paper and rolled onto a heated mandrel, with a weighted head roller providing the pressure. Tubes are also made with a glass fabric reinforcement, usually with polyester or epoxy resin.

Many applications for glass filament wound tubes and rings are being found. The rings are made by winding the glass filaments into a mould. The tubes are wound by laying a band of rovings about 25 mm wide onto a mandrel as a helix, in both directions, the second complete pass just touching the first. The angle of winding can be varied, depending on if the requirement is for high axial strength or high bursting strength. The resin system is usually epoxide, but may be polyester.

The applications for tubes may be as the coil supports for power transformers, as transformer tap changer housings, and for switchgear cross jet pots. Filament wound rings are used to support the overhang of rotating machine stator windings.

A range of flexible composites are made for rotating machine slot insulation, often combining a mechanically stiffer product to support a very flexible material of high electric strength. The materials are usually bonded with a thermosetting adhesive.

Examples are polyethylene terephthalate film (e.g. Melinex - ICI or Mylar - Du Pont) combined with paper, presspaper or polyester fibre mat of various thickness, recommended for use in machines up to Class B rating although the use of the polyester fibre mat extends the rating towards Class F.

For higher temperature ratings, a combination with a high temperature polyamide paper (e.g. Nomex - Du Pont) is used, such as the polyethylene terephthalate film for Class F machines and polyamide film (e.g. Kapton - Du Pont) for Class H and Class C machines.

(For further reading, see reference 8.2).

8.4   CASTINGS AND MOULDINGS

The distinction between a casting and a moulding, is that the former is produced by pouring the liquid resin system into a mould and the latter requires that the resin system is forced into the mould under pressure.

For electrical insulation components for high voltage applications, it is necessary for the material to be either void free, or if any voids exist, that they do not discharge in service. When the casting method of production is used, this usually requires a vacuum process at the mixing stage and sometimes at the pouring stage.

Castings are produced from epoxide resin systems, the filler being necessary to reduce the cost and the shrinkage of the component. The most common filler is silica flour, but this is not acceptable in some applications, (for example, for use in sulphur hexafluoride) and alternatives such as bauxite, alumina and dolomite are used. A marked reduction in some properties results as shown in Table 8.2.

TABLE 8.2   Typical properties of cast epoxide resin with various fillers

Property	BS 2782 Test Method	Filler			
		Silica	Bauxite	Alumina	Dolomite
S.G.	620A	1.8	2.15	2.35	1.8
H.D.T. ^{o}C	121A	110	105	96	105
Crossbreaking MN/m^2	335A	96.5	85	69.5	71.5
Crushing MN/m^2	345A	186	186	141	140
Shear MN/m^2	340A	54	51	48	52
E.S. KV/m^2	201C	16	15	13.5	14.7

The original cast epoxide process, still in use, is to pour the resin system into the mould in an autoclave; the filled moulds are then removed to a curing oven. Very large castings are made by this process.

Considerable advances in curing techniques have been made in recent years, which requires electrical heaters to be inserted in the moulds in positions which control the cure.

By such techniques, curing times have been reduced considerably and although the moulds are more expensive to make, they are justified if greater output is required. Automatic mould handling equipment has also contributed to a quicker turn round of the moulds, making the process less labour intensive. The process has advanced to the stage that multi part tools opening in different directions allow quite complicated mouldings to be produced.

More recent developments have produced a range of polyurethane casting resins filled with magnesium/calcium carbonate. The resin is supplied with the fillers already mixed in, requiring only the addition of the hardener. One limitation in the processing of this system is that the higher the glass transition temperature (Tg), the lower is the pot life, the material with the highest Tg quoted by the manufacturers having a pot life of only 10 to 15 minutes (Ref 8.3). This makes vacuum pouring difficult and the material is usually poured without vacuum the resin having been prepared under a vacuum process by the manufacturers.

Compared with the conventional cast epoxide resin process, it is claimed by moulders of polyurethane, that the cost reduction may be as much as 40%. However, this difference has been almost entirely eroded by the use of epoxide moulding powders, although the tool cost for polyurethane moulding would be less, hence reducing the initial investment.

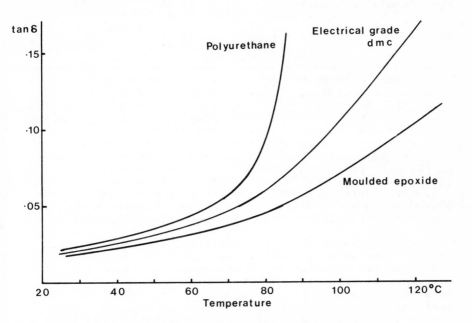

Fig.8.3   The relation between loss tangent and temperature for three moulded bushings

Polyurethane is being moulded for some electrical applications but the grade normally selected as having a heat distortion point comparable with cast resin, has been found to have inferior mechanical and electrical properties. In Fig.8.3 are given the loss tangent/temperature characteristics of identical bushings produced from moulded epoxide resin, polyester/dough moulding compound (d.m.c.) and polyurethane, from which it is seen that the knee of the curve for polyurethane occurs at a lower temperature than either the d.m.c. or the epoxide. Identical post insulators with moulded in inserts in both faces, produced from both cast epoxide and polyurethane were subjected to tensile and cantilever tests. Through bushings made from cast epoxide or polyurethane were subjected to a conductor push out test and a cantilever tests. The mean results are given in Table 8.3

TABLE 8.3   Comparison of properties of components made from cast epoxide or polyurethane

Type of Moulding	Test	Cast Resin	Polyurethane
Post Insulator	Tensile	24 KN	15 KN
	Cantilever	6 KN	3 KN
Bushing	Conductor push out	44 KN	26 KN
	Cantilever	6 KN	6 KN

Three moulding methods are in general production, compression moulding, transfer moulding and injection moulding. All require high quality matched moulds, usually made in two parts, one of which remains stationary and the other is used to open and close the mould.

The basic difference in the technique, is that for a press moulding, the charge is loaded between the two parts of the open pre-heated mould. A transfer moulding, is made by loading the charge into a cavity adjacent to the component cavity where it is softened and forced into the closed component cavity. An injection moulding draws its charge from a hopper by a feed screw principle, or plunger, also into the closed mould cavity.

With these mouldings, it is to be expected that a "flash" line will be apparent on the surface of the moulding where the two parts of the mould meet and either a riser or a sprue will need to be removed which will leave a small area differing slightly in appearance from the as moulded surface.

Mouldings of all three types are removed from the mould fully cured and the processes all allow the moulding in of conductors and inserts. Moulding pressures, temperatures and times vary considerable for the resin system and thickness of component, so that the following figures should only be regarded as typical.

For polyester dough moulding compound (d.m.c.) and sheet moulding compound (s.m.c.) curing temperatures in the range $130^{\circ} - 150^{\circ}C$, are used. Pressures are of the order of 1.5 to 5 MN/m^2 for d.m.c. and 2.75 to 7 MN/m^2 for s.m.c. and the curing time would be expected to be 60 to 90 seconds for a thickness of 3 mm.

There is a considerable range of d.m.c. and s.m.c. compounds available and the application and requirements need to be discussed with the moulding manufacturer, in order to obtain the most suitable grade. The s.m.c.'s tend to give higher mechanical properties because they can be produced with longer fibre lengths - up to 50 mm, and the fibres are not degraded as they could be with d.m.c. if sufficient care is not taken at the mixing stage. Degradation of the glass fibre in a d.m.c. can occur in an injection moulding machine, resulting in a reduction of impact strength of more than 50% for 12 mm length fibre.

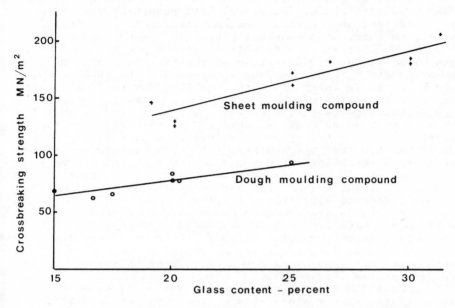

Fig.8.4   The effect of glass content on the crossbreaking strength of d.m.c. and s.m.c.

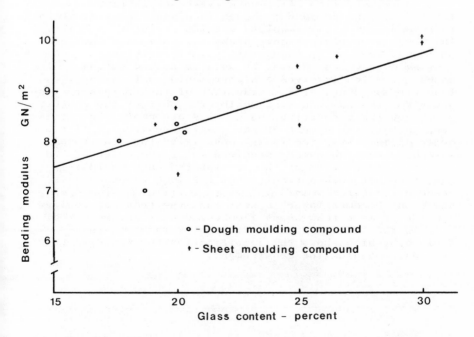

Fig.8.5   The effect of glass content on the bending
modulus of d.m.c. and s.m.c.

The glass content can be varied over a considerable range for both d.m.c. and s.m.c. with resulting variation in mechanical properties. For any glass content, there will be a range of mechanical properties depending on fibre length, so that it is difficult to give a precise picture graphically, but an indication of the variation of cross-breaking strength and bending modulus is given in Figs. 8.4 and 8.5. It is interesting to note that whereas the ultimate strengths of the d.m.c. and the s.m.c. produce two distinct characteristics with different slopes, the results of bending modulus, obtained at much lower stress levels, from the two types of material, cannot be distinguished from one another. The reduction in the slope of the crossbreaking curve for d.m.c. may be attributed to the break up of the fibres during mixing. The variations in the results for similar glass contents may be accounted for by different glass fibre lengths.

Some grades are produced which have good arc resistance, high fire retardancy and good electrical properties at temperatures up to 70° - 80°C. Generally, the knee in the loss tangent/temperature curve occurs at this order of temperature compared with perhaps 120° for an epoxide system (Fig. 8.3).

Some glass reinforced polyester materials tend to absorb water more rapidly than do the epoxides, so that under damp conditions, the electrical properties are likely to degrade more rapidly.

An illustration of comparative costs is given in Table 8.4. The component chosen is a small bushing which is to be made in cast or moulded epoxide resin or in d.m.c. The first line of the table shows the estimated cost of the mould to produce the component by the traditional vacuum casting oven cured method, by an accelerated curing technique which involves a higher mould cost, or by transfer moulding. The second line of the table shows the price for the component, ignoring the cost of the moulds.

When the respective mould costs are amortized, it is seen that if the number of components required is of the order of 100, then the lowest cost is by the traditonal method. When the number required is of the order of 500, then the higher cost of the accelerated cure mould is justified but when a requirement for 1000 components is reached, transfer moulding is more economic. The d.m.c. component becomes lower in cost than the transfer moulded epoxide if more than about 1500 components are required.

An indication of production rates is given in Table 8.5 which shows the justification for the different methods, based on the requirement.

TABLE 8.4   Price compaison for an identical component
made by different production methods

		Epoxide		
	Traditional	Fast Cure	Transfer Moulded	d.m.c.
Tool Cost	£800	£1300	£2800	£3500
Component Price	£5.49	£3.84	£2.21	£1.61
Total Price each/100	£13.49	£16.84	£30.21	£36.61
Total Price each/500	£7.09	£6.96	£7.81	£8.61
Total Price each/1000	£6.29	£5.14	£5.01	£5.11
Total Price each/5000	£5.65	£4.10	£2.77	£2.31
Total Price each/10,000	£5.57	£3.97	£2.49	£1.96

TABLE 8.5   Comparison of production rates for an identical
component made by different production methods.

		Epoxide		
	Traditional	Fast Cure	Transfer Moulded	d.m.c.
Components produced per week	5	30	800	1000
Weeks to produce 1000	200	33.3	1.25	1.0

8.5   EXTRUSIONS AND PULTRUSION

The extrusion process has been used extensively for
the forming of thermoplastics. The polymer is drawn from
a feed hopper, usually in granular form, melted, compacted
and forced through a die to obtain the final form. Very
thin film, sheet, tubing and structural sections may be
made by this process.
The pultrusion is used when a reinforcement, strong
enough to be pulled through a die, is a constituent part of
the product. A mixture of reinforcements, usually glass,
may be used such as rovings, mat and fabric. The
reinforcements are pulled into an impregnation chamber and
then into a heated die, the material being fully cured

when it emerges from the die. Polyester, epoxide and
acrylic are the usual resin systems. The polyester
materials are usually the lowest in cost, the epoxides tend
to give the best mechanical properties and the acrylic,
the highest thermal performance. The mechanical properties
depend very much on the configuration of the reinforcement,
but one manufacturer claims a tensile strength of 1.4 GN/m^2
with uni-directional glass reinforcement and epoxide resin.
With a mixed reinforcement, 400 MN/m^2 would be typical.

8.6   MATERIAL SELECTION

Although cost is very important, the first
consideration made by many insulation engineers when
selecting a material, is to decide if it will stand the
expected working temperatures. Unless space is restricted,
it is usually possible to use sufficient material to
withstand the electrical or mechanical stresses, but if it
softens at the working temperature, looses too much strength
or ages rapidly, then there is little that can be done about
it.
There are a number of evaluations which can be made in
order to assess the effect of temperature. The Heat
Distortion Temperature (HDT) is favoured for comparison
purposes. The usual method of test is to increase the
temperature at a constant rate under a constant load and
to measure the temperature at which a pre-determined
deflection occurs.
Alternatively, mechanical or electrical properties
over a range of temperature may be measured.
Characteristics, of the type shown in Fig. 8.6 are obtained,
from which a good indication of the suitability of the
material for use at a particular temperature, may be
indicated.
For power equipment, which can be expected to operate
for many years, long term aging data is essential. This
may be obtained in accordance with IEC Recommendation 216-
Guide for the determination of thermal endurance properties
of electrical insulating materials. Aging processes are
the subject of another chapter but the value of such testing
is illustrated in Fig. 8.7, in which a comparison is made
between two glass reinforced laminates, the IEC 216 method
having been extended to provide data relative to working
temperatures. From this data, it can be seen that if the
glass mat reinforced epoxide laminate was operated
continuously at $155^{\circ}C$, it could be expected to have a 50%
strength retention after more than 20 years. However, it
will be seen that to obtain such data, aging times of up to
20,000 hours, that is, more than 2 years, are required in
order to comply with the constraints of the specification.
How then is a solid insulating material selected? It
has been attempted to show that there are three primary
considerations, the working temperature range, the method
of manufacture and closely connected with both of these,
the cost. Often, features such as electrical and
mechanical characteristics, water absorption, arc

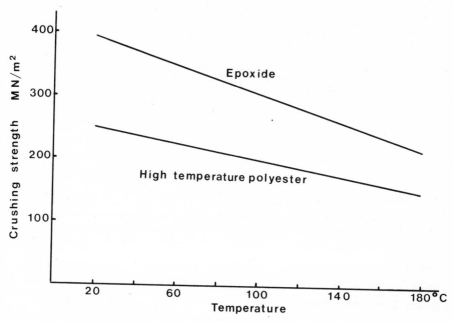

Fig.8.6   The effect of temperature on the crushing
strength of glass reinforced laminates

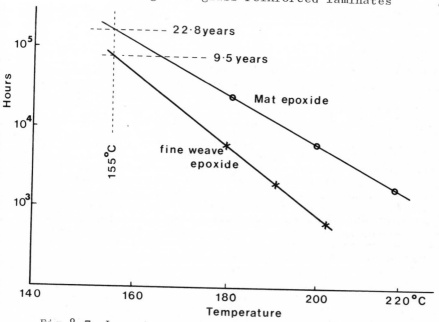

Fig.8.7   Long term aging curves for glass reinforced
laminates tested at 155°C, used to indicate
life at 155°C

resistance and flame retardancy are properties which may be considered, depending on the application, once it is established that the material will meet these first three requirements.

TABLE 8.6 Typical short term and long term working
temperatures for laminates

Note A. Lower for thickness above 6 mm

Note B. Depends on type of polyester resin

Resin/Reinforcement	Working Temperatures C^o	
	Short Term	Long Term
Phenolic/Paper	120	90
Phenolic/Wood Veneer	120	100
Phenolic/Asbestos Paper	185 (A)	165 (A)
Phenolic/Asbestos Fabric	185 (A)	165 (A)
Phenolic/Cotton Fabric	130	120
Phenolic/Glass	155	140
Epoxide/Paper	140	130
Epoxide/Glass	180	155
Polyimide/Glass	> 300	160
Epoxide/Cotton	150	130
Silicone/Glass	230	190
Polyester/Glass	155-180 (B)	140-170 (B)
Silicone Asbestos	220	180

Generally, the higher the temperature classification of the material, the higher is the material cost. An indication of typical working temperatures is given in Table 8.6 from which it is seen that working temperature cannot be considered in isolation from operating life. Manufacturers would probably consider short term life to be 5000 - 10,000 hours and long term life to be 10-30 years.

Strength retention at the working temperature must also be considered and Table 8.7 shows what may be expected of a range of laminates.

TABLE 8.7   Typical retention of crossbreaking strength of laminates at high temperatures

Material	Percentage strength retention		
	$20^{\circ}/155^{\circ}$C	$20^{\circ}/180^{\circ}$C	$20^{\circ}/250^{\circ}$C
Glass Fabric/Phenolic	60		
Glass Fabric/Epoxide	75	60	
Glass Fabric/Polyimide	90		77
Glass Fabric/Silicone	65	55	45
Glass Mat/Epoxide	75	60	
Asbestos Fabric/Phenolic	70		

Little space has been given to thermoplastics. Care needs to be taken when selecting thermoplastics for use at high temperature because as the name implies they will soften. However, some are suitable for use at quite high temperatures, polytetra-fluoroethylene for example. There is a tendancy for many thermoplastics to creep dimensionally at modest working temperatures.

The choice of manufacturing method depends on the number of components required, their shape and the addition of metalwork; rotating machine slot wedges, for example, will probably be machined from a flat sheet or pultruded. Although bushings are still made from resin bonded paper, at the lower voltage range, casting and moulding methods are used, which allows a flange with inserts to be provided with no extra assembly time and accomodates the large quantities usually required.

However, most manufactures of solid insulating materials offer a design service and will be ready to discuss the most suitable material and the most economic method of manufacture for the application.

8.7   ACKNOWLEDGEMENTS

During the preparation of this chapter, discussions with members of the staff of the following companies who have provided data, is gratefully acknowledged.

BTR-Permali RP Ltd
Ciba Geigy Ltd
Dow Corning
Permali Gloucester Ltd
H.D. Symons and Co Ltd
T.A.C. Construction Materials Ltd
Tufnol Ltd

REFERENCES

8.1   Billmeyer, F.W.,
      Textbook of polymer science. John Willey and Sons,
      New York.

8.2   Morgan, P, (Ed).,
      Glass reinforced plastics. Iliffe Books Ltd, London,
      England.

8.3   Lottanti, G, and Schiegg K., 1976,
      Electrical engineering applications of polyurethane
      resins. Kunststaffe - Plastics 23.

8.4   Davis, J.H., 1959,
      Silicone electrical insulation. Proc. IEE Vol. 106
      Part A.

8.5   Neal, J.E., and Whitman, A.G., 1982,
      Insulation of high voltage rotating machines.
      IEE Conference Publication 213, pp 23-28.

8.6   Heighes, J, 1982
      The provision of design data from extended heat
      aging tests. IEE Conference Publication 123, pp 41-45

8.7   Groves, D.J., 1979,
      An order of merit for insulation. IEE Conference
      Publication 177, pp 157-160

8.8   Aggleton, M.J., 1977,
      Non-woven glass laminates. Circuit World Vol. 4
      No. 1.

8.9   Kyoichi Shibayama, 1975,
      Temperature dependence of the physical properties
      of crosslinked polymers. Progress in Organic Coatings,
      3 pp 245-260

8.10 Gaggar, S.K., and Broutman, L.J., 1976, .
      Effect of matrix ductility and interface treatment
      on mechanical properties of glass fiber mat composites.
      Polymer Engineering and Science, Vol. 16 No. 8.

8.11 Maries, K,
      Prediction of thermal conductivity of g.r.p. laminates.
      Fire Research Station report, CP 70/76.

Chapter 9

Properties and applications of solid/liquid composites

J. Staight

9.1 INTRODUCTION

In the early part of this century workers added oil to cellulose insulation probably in an attempt to exclude moisture and this combination was soon accepted as an exceptionally good form of high voltage insulation. Gradually as the mechanism of dielectric failure associated with partial discharges became better understood, the development and use of composite systems was put on a firm basis.

The use of oil impregnated cellulose in high voltage equipment has persisted to the present day despite the competition from a variety of synthetic materials. If we make a simple comparison of the electrical characteristics of a synthetic polymer such as polypropylene with insulation paper, we discover that the synthetic material gives the advantages of lower loss angle, lower moisture absorption and increased electric strength. Notwithstanding this the replacement of paper by film in an engineering context is far from straightforward. Physical factors governing the level of impregnation and the mechanical strength of the composite are just two of the important considerations.

In general choosing compatible materials for a given application requires information on the appropriate physical and chemical interactions in addition to a knowledge of their electrical characteristics. The present chapter deals with the use in composite form of some common solid and liquid dielectrics and their application to power equipment. Attention is centred on the use of polymeric materials which are currently available in the form of sheet or tape and which are impregnated with oil or synthetic fluid.

9.2 IMPORTANT PHYSICAL PROPERTIES OF MATERIALS

Insulation may be in the form of a single sheet but more often it consists of multiple layers between which the impregnant must be introduced. The physical characteristics of the solid and liquid determine the speed and efficiency of the impregnation process. When complete the insulation will have to fulfil a mechanical as well as an electrical role and again physical factors are responsible for determining the strength and stability of the composite structure.

9.2.1   Solid Insulation

An open structure which is required for efficient
impregnation is readily obtained by the use of materials
with textured surfaces.   Paper has a natural surface texture
which is dependent on the extent of the calendering process
and hence the final density of the product.   Little difficulty is experienced in vacuum impregnating papers of the
highest available density.   In contrast to this most synthetic polymers are initially smooth at the point of extrusion or casting and special techniques are required to
create surface texture.   The extent to which this is carried
out with polypropylene film is shown in Fig. 9.1.   In this
electron micrograph the surface structure of capacitor grade
film on the left can be compared with a medium density paper
on the right.   Other types of polymer are not generally
available with such highly treated surfaces.

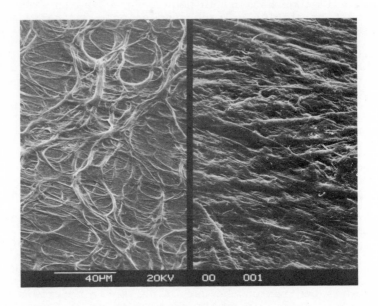

Fig.9.1   The surface topography of medium
          density capacitor paper (right)
          and textured polypropylene (left)

Apart from allowing fluid to penetrate the insulation
the open structure is required to permit a thorough degassing and drying of the insulation before impregnation.   This
is very important in the case of cellulose because the water
content is initially very high and a small residual amount
left in the insulation effects both the electrical properties and the resistance to thermal degradation.   In the case
of cellulose vacuum treatment at elevated temperature
usually in the range 105 to 120°C is advocated.

With cellulosic insulation thermal degradation is present during processing and throughout its operating life. H F Church collated information covering the effect of temperature on ageing using a number of different criteria to determine life. The results which are summarised in Fig. 9.2, relate mean 'life' to temperature. The numbers adjacent to the data indicate the sources of information given in Ref.9.1. In Fig. 9.2 the slope shown represents the activation energy of the ageing process in accordance with the Arrhenius equation.

TABLE 9.1.   The heat distortion temperature (19.2 kg/cm^2 ASTM D648) of some engineering plastics

Polymer	Specific gravity	Heat distortion temperature, $^\circ$C
Polyethylene (low density)	0.92	37
Polyethylene (high density)	0.95	46
Polypropylene	0.91	60
Polycarbonate	1.20	130
Polysulphone	1.24	174
Polyphenylene oxide	1.06	191

Certain mechanical properties of insulation such as tensile strength may be important during the fabrication of equipment whereas others such as creep assume greater importance during its operating life. A disadvantage of some synthetic polymers is the poor load bearing capability which derives from the occurrence of plastic flow. Table 9.1 gives the heat distortion temperatures of some typical engineering plastics. The values given are for non-oriented polymer. In practice many electrical grade films are biaxially oriented and receive special crystallisation treatment, which can have a marked effect on their mechanical and thermal properties. Oriented films are reasonably stable within their operating temperature range but they are often subject to a fairly rapid shrinkage in area above a critical temperature. Polymers differ in their ability to crystallise in general, macromolecules with relatively rigid and inflexible chains crystallise most easily. This rigidity can result from rigid chain links such as the anhydroglucose rings in cellulose or the benzene rings in polyethylene terephthalate (PET). It is found that insulation based on natural fibre usually contains sufficient crystallinity to remain stable at comparatively high mechanical load.

9.2.2   Impregnants.

There are various dielectric fluids which are currently in use as insulation impregnants, these range from low cost

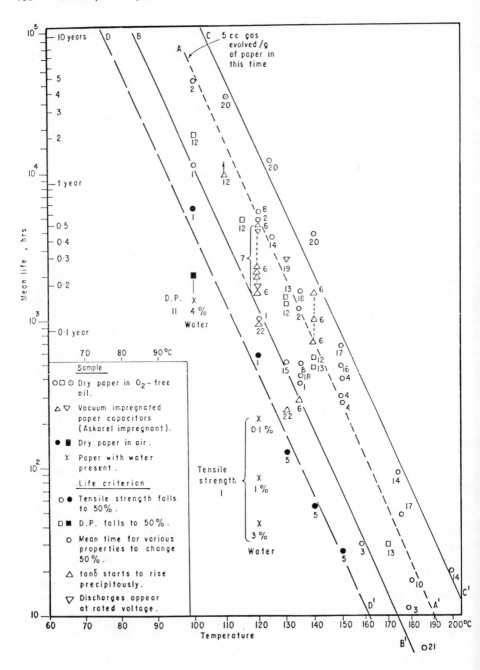

Fig.9.2   Degradation of paper and impregnated-
paper at various temperatures.
Without applied stress

petroleum oils to specially synthesised products, and have
been fully described in Chapter 7. The fundamental physical
requirement is for fluid flow over the operating temperature
range, thus the viscosity versus temperature relationship is
of special interest. For Newtonian fluids this relationship
is predictable down to within a few degrees of the pour
point of the fluid. Outdoor power equipment is required to
function over a wide temperature range and for operation in
the lower temperature categories the crystallisation temperature and low temperature viscosity of the fluid must be
considered. Table 9.2 gives some characteristics of various
impregnants including the pour point and boiling point. The
latter is not likely to be exceeded in service except under
fault conditions but the vapour pressure during vacuum
impregnation is relevant.

TABLE 9.2. Properties of various impregnants

	Specific gravity	Permitti- vity 50 Hz 20°C	Viscosity 20°C Cp	Pour point °C	Boiling point or range °C
Mineral oil	0.87	2.2	12	-50	Indefinite
Trichloro- diphenyl	1.38	5.9	80	-19	325-360
Monoisopro- pylbiphenyl	0.99	2.5	8	-50	295-300
Dodecylben- zene	0.87	2.3	10	-50	277-296
Polybutene	0.85	2.1	32	-50	Indefinite
Dioctyl- phthalate	0.99	5.3	81	-45	386
Silicone oil	0.96	2.7	22	-60	320
Mono/dibenzyl- toluene	1.01	2.7	7	-50	300-310
Castor oil	0.96	4.5	800	-18	Indefinite

High stability is required from high voltage insulation and impregnants should exhibit a high resistance to
thermal degradation and in particular to oxidation. Until
recently chlorinated impregnants such as trichlorodiphenyl
were widely used because of their stability and fire resistance. Unfortunately the stability of these fluids also
results in a high persistence in the environment which is
undesirable. The use of these fluids in composite insulation has been curtailed in favour of various synthetic
fluids which include organic esters and silicones. Various
aromatic hydrocarbons are finding increasing use in the more
highly stressed systems. If necessary their oxidation
resistance can be improved by the use of inhibitors.

9.2.3 Liquid-Solid Interaction

Impregnation can be regarded as a physical interactive
process controlled predominantly by the surface tension

force. The rate determining factors include the viscosity
of the impregnant, its surface tension and the contact angle
with the dielectric. A low contact angle (θ) between the
liquid and solid favours rapid penetration of the liquid
between the insulation surfaces and into the hollows or
pores in the surface. The surface tension force is basic-
ally $\sigma \cos \theta$, where σ is the surface tension of the liquid,
therefore it is very dependent on the nature of the impreg-
nant. As an example, the contact angles between polypropylene dielectric and a number of impregnants are given in
Fig. 9.3 which is derived from Ref. 9.2. Perfectly wetting
liquids are generally non-polar whereas highly polar liquids
tend to have high contact angles. The results show that
among the polar liquids of similar permittivity there are
considerable differences in contact angle. In a particular
system the contact angle can be altered by blending together
different fluids or by altering the surface condition of the
solid, for example, corona treatment of a polymer surface
during manufacture can reduce the contact angle with polar
impregnants.

Most of the commonly used impregnants are absorbed to
some extent by polymeric insulation and this interaction can
have a profound effect on the electrical and physical
characteristics of the composite insulation. Fendley and
Parkman (Ref. 9.2) show that the electric strength of poly-
olefins is improved by impregnation presumably due to the
effect of impregnation on physical defects in the polymer.
Work carried out on polypropylene film containing observable
defects in the form of small cavities has shown that the
time (τ) required to fill such cavities is linearly related
to the cavity size and very dependent on the nature of the
impregnant as shown in Fig. 9.4.

Excessive fluid absorption by the dielectric results in
swelling which produces several disadvantages. Full impregnation may even be prevented if there is undue swelling in
the outer layers of the insulation. The extent and time-
scale for the swelling of polypropylene is given in Fig. 9.5
for a wide range of fluids.

The mechanical properties of the film can also be
affected. Reductions in tensile strength have been noted
for polypropylene aged at elevated temperature in the presence of several impregnants (Ref. 9.4) depending on the
crystallinity and orientation of the film.

With glassy polymers such as polyphenylene oxide, poly-
carbonate and polysulphone the absorption of impregnant can
cause a reduction in the glass transition temperature. This
is accompanied by an increased risk of crazing and stress
cracking which can severely limit the use of such materials
as high voltage insulation.

Chemical interaction between solid and liquid dielectrics is generally confined to various leaching processes
whereby low molecular weight or ionic components migrate
from the solid to liquid or vice versa. The presence of
ionic contamination in the liquid can be detected by an
increase in conductivity or loss angle and in capacitors
there may be a noticeable greater Garton effect (Ref. 9.5)

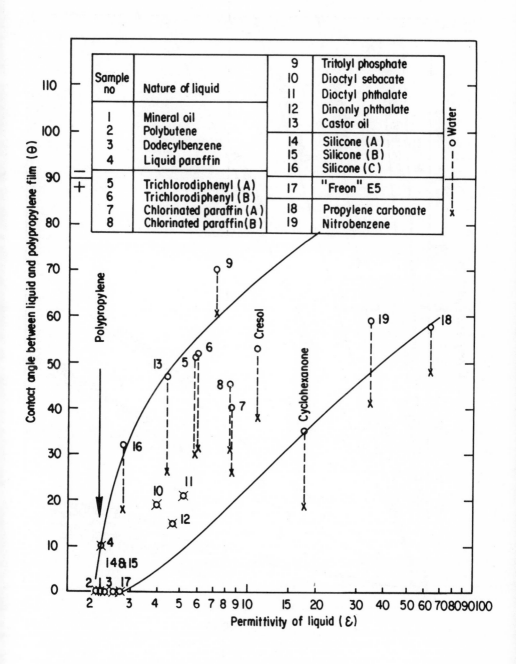

Fig.9.3 Contact angle between polypropylene
film and various liquids in air v.
ε of liquid

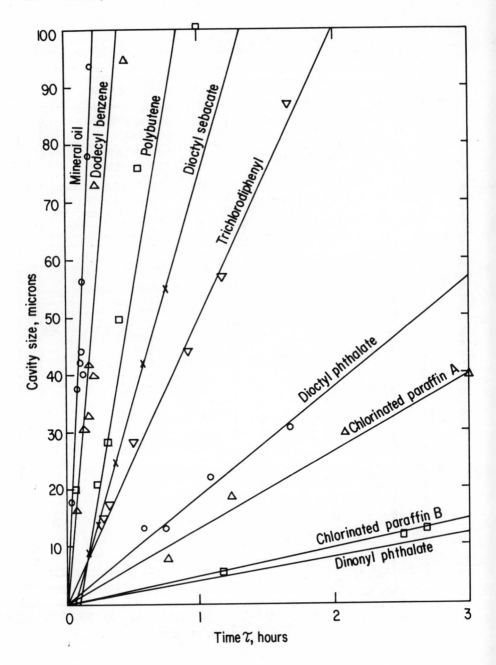

Fig.9.4   Dependences of time (τ) required
to fill a cavity in polypropylene
film on the initial size of the
cavity for different liquids

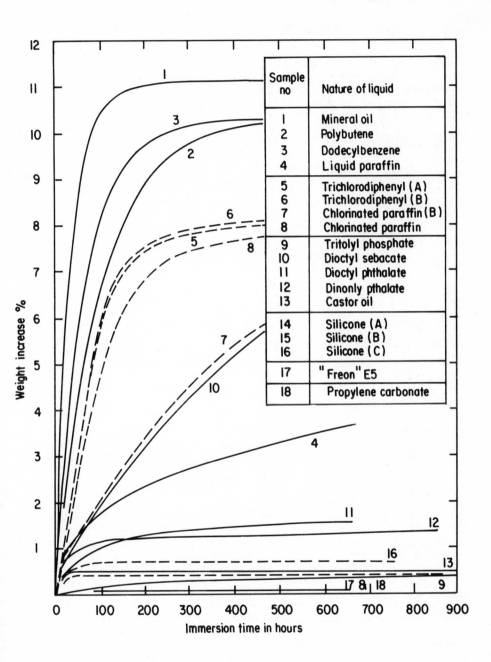

Sample no	Nature of liquid
1	Mineral oil
2	Polybutene
3	Dodecylbenzene
4	Liquid paraffin
5	Trichlorodiphenyl (A)
6	Trichlorodiphenyl (B)
7	Chlorinated paraffin (B)
8	Chlorinated paraffin
9	Tritolyl phosphate
10	Dioctyl sebacate
11	Dioctyl phthalate
12	Dinonly pthalate
13	Castor oil
14	Silicone (A)
15	Silicone (B)
16	Silicone (C)
17	"Freon" E5
18	Propylene carbonate

Fig.9.5   Weight increase of polypropylene
film during immersion in various
liquids at room temperature ($\sim 20^{\circ}$C)

at low stress. The presence of ionic contaminants can lead
to electrochemical failures of the dielectric during service
and for this reason stringent precautions are taken to eliminate contaminants during the manufacture of dielectric
materials. Ions may be immobilised under service conditions
for example, by including an active filler in a cellulose
dielectric or by the addition of special scavenging chemicals to the impregnant.

9.3   ELECTRICAL CHARACTERISTICS

In power system equipment an insulation life greater
than 20 years is required under conditions where the insulation may experience transient voltages well in excess of the
nominal operating value. At the manufacturing stage equipment is checked using short term tests based on impulse and
other transient waveforms. Discharge inception voltages are
monitored and prospective long term performance assessed
either from field experience or from the results of accelerated endurance tests.

The electric strength of the insulation is clearly an
important design parameter and for composite systems note
has to be taken of the various breakdown processes which are
described in Chapter 4. These emphasise the need for
insulation which is free from voids and contaminants of an
ionic or particulate nature. In this section consideration
is given to other electrical characteristics, namely the
effects of permittivity, loss-angle and the field distribution in the dielectric.

9.3.1   Permittivity and Loss-Angle

Permittivities of the commonly used impregnated polymers range from about 2 to 6. The upper values are for
cellulosic insulation impregnated with polar liquids such as
trichlorodiphenyl and dioctylphthalate. A high permittivity
is disadvantageous in certain applications such as high
voltage cable insulation because of the high charging current. In capacitors a high permittivity dielectric may give
an improved volume efficiency but this is very dependent on
the working stress. The reactive power (kvar) of a capacitor considered on a unit volume basis is proportional to
dielectric permittivity and to the square of the operating
stress. In commercial practice considerably higher operating stresses are possible with a synthetic polymer dielectric such as polypropylene than with paper. One reason for
this is the high incidence of weak spots in paper compared
with most synthetic dielectrics. Evidence of this is given
in Fig. 9.6 which shows the number of breakdowns per unit
area at a given stress for single layers of dry paper and
film passed between mercury test electrodes.

The loss-angle of high voltage insulation is important
because a high value results in thermal limitations on a.c.
equipment designs. For example the insulation stress may
have to be reduced to prevent a thermal runaway condition.
Particular care is required when the loss-angle exhibits a

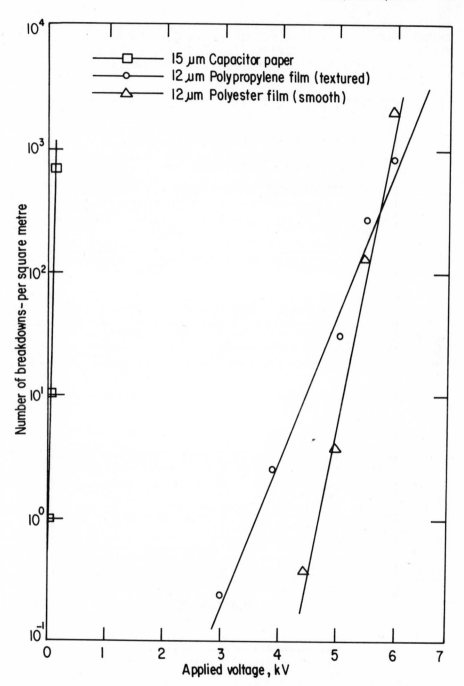

Fig.9.6   Defects detected using mercury
          electrode test in capacitor
          tissue and in various films

rapid increase with temperature as shown for polyethylene-terephthalate in Fig. 9.7. In capacitor technology the replacement of paper dielectric with polypropylene has resulted in units with a low energy consumption and negligible self-heating at power frequency.

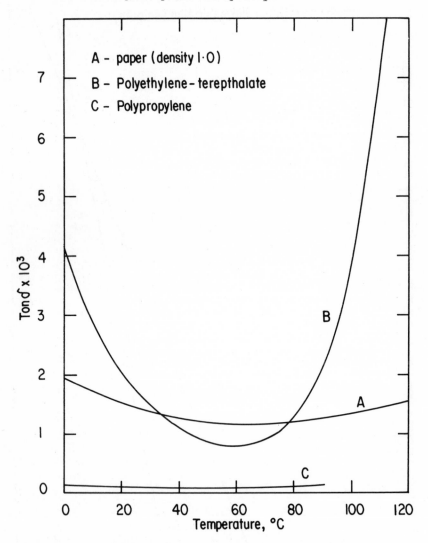

Fig. 9.7   Loss tangent as a function of temperature for dry paper and various synthetic polymers

9.3.2 Stress Distribution

In using composite insulation a knowledge of the spatial stress distribution and the division of stress between

the different components is important. With d.c. applied
the stress is divided resistively, this is not a straightforward process because resistivity of the dielectric is
stress, time and temperature dependent. The effect of this
is seen with impregnated paper cables where a reversal of
the stress distribution is found between the on and off load
condition. In a multiple array such as a high voltage capacitor additional fixed resistors may be added for stress
equalisation.

     The majority of power applications involve a.c. stress
where the stress is based largely on elemental capacitance.
That is to say in a uniform field, the stress is inversely
proportional to the local permittivity. Because both the
system geometry and permittivity are fairly stable parameters, the stress distribution in an a.c. system is well
defined. In certain cases resistive components may be added
for stress control, an example is the use of semi-conducting
tapes in the manufacture of high voltage cables.

     In designing equipment attention must be given to avoid-
ing high local stress and it must be remembered that stress
enhancement will occur at any conductor edge with a sharp
profile which is in contact with the stressed insulation.
Partial breakdown of the insulation at regions of high field
is particularly damaging in impregnated insulation because
gas bubbles tend to form which produce a further reduction
in electric strength. For this reason highly stressed components such as power capacitors usually contain impregnants
which are very gas absorbing. When cellulose insulation is
involved it has been shown by Krasucki (Ref.9.5) that bubble
formation under high field conditions is promoted by the
presence of absorbed moisture and this can be related directly to the discharge inception stress. Measurements on
samples of impregnated paper dried to different degrees and
immersed in impregnant show that discharge inception stress
measured at test electrodes increases with increasing
dryness.

9.4   THE APPLICATION OF SOLID/LIQUID COMPOSITES TO POWER EQUIPMENT

9.4.1 Cables

     Oil-filled cables are used for power distribution at
medium and high voltage. In the pressurised type the volume
changes which occur due to alteration in temperature are com-
pensated by the movement of impregnant to and from a collaps-
ible reservoir. A mobile impregnant is used and the oil
moves freely along special ducts which are situated in zero
or low field regions. Single core cables are usually manufactured with hollow conductors but in high voltage 3 core
cables separate oil ducts may be provided. In Britain it is
customary to allow for a static pressure of about 500 kN/m^2.
The advantage of the pressurised system is that there is a
greatly reduced risk of degradation by partial discharges in
the insulation and the operating stress can be comparably
high. Commercial oil impregnated cables are currently

provided with paper dielectric and for a.c. application at
275 or 400 kV(rms), maximum stresses up to 13 kV(rms)/mm are
acceptable. For the bulk transmission of power over long
distances it is feasible to use HVDC. In d.c. cables a more
efficient use can be made of the insulation giving an average operating stress of up to 20 kV/mm. Under d.c. conditions the effect of charging current and loss angle are minimised enabling a high density paper to be used.

In high voltage cables the dielectric thickness is generally determined by its ability to withstand impulse breakdown. It is found that the a.c. and impulse electric
strength is dependent on paper thickness. For this reason a
graded construction may be used with thin layers of paper
near to the conductor.

Cable impregnants are typically refined mineral oil or
alkylbenzene. The presence of an aromatic component provides the added safeguard of high gas absorption. The
impregnant is usually formulated to give a reasonable compromise between gas absorption and oxidation resistance.
The current rating of a cable is dictated by thermal considerations and it is usual to limit the maximum continuous
conductor temperature to about 85°C. Some heat is contributed by the dielectric losses and at high voltage the effect of this is significant. The dielectric loss of a cable
with a capacitance C per unit length and a dielectric loss
tangent of tan δ is given by:

$$p = 2 \pi f C V^2 \tan \delta \dots\dots\dots\dots\dots\dots (9.1)$$

where p is the loss per unit length at voltage V and frequency f.

The effect of this loss on dielectric temperature and
power dissipation is to reduce the rated current of the
cable. At 33 kV(rms) this effect is negligible but at
400 kV(rms) the figure is about 15%. Clearly it would be of
benefit to use a low loss dielectric at high voltages and
the application of synthetic polymers to such cables is being actively researched. Edwards and Melville (Ref.9.7)
discuss the use of a laminate comprising a layer of polypropylene sandwiched between two layers of paper. In this work
the construction of cables with significantly improved loss
angle is shown to be feasible. Another approach is to use a
synthetic dielectric composed of a glassy polymer such as
polyphenylene oxide (Ref.9.8), however the stress cracking
problem experienced with this class of insulation has held
up the development of a practical system.

9.4.2 Capacitors

High voltage capacitors are found in precision devices
such as voltage dividers and in high kvar banks for power-
factor correction on inductively loaded circuits. The
active dielectric is invariably an impregnated polymer of
natural or synthetic origin. Divider systems generally contain a dielectric of mineral oil impregnated paper but for
power-factor correction capacitors the trend is towards an

increasing use of polypropylene film. The lower loss angle
obtained with polypropylene gives a reduction in dielectric
heating and lower power consumption both of which are advantageous where large volumes of dielectric are involved.
Power factor correction capacitors are produced either with
a mixture of paper and film or with all-film dielectric.
The latest designs have loss tangent which is only one tenth
of the value for the early paper types.
      Capacitors contain highly stressed dielectric and the
choice of impregnant is particularly important. Usually
highly aromatic and hence gas absorbing fluids are chosen.
Until recently polychloronatedbiphenyls were used widely as
capacitor impregnants but their use has now been replaced by
non-chlorinated fluids such as monoisopropylbiphenyl.
      The development of specially textured films and new
fluids has allowed operating stresses to be progressively
increased from 18 kV(rms)/mm applicable to paper capacitors
to about 55 kV(rms)/mm for the all-film type. Capacitors
are insensitive to very short impulses and the dielectric
thickness is usually determined by the rated operating
stress. There has been some concern expressed over the
effects of transient over-voltages which persist for several
cycles at power frequency and new tests have been proposed
which take this into account. In commercial capacitors the
upper limit on operating voltage is set by the discharge
inception voltage, this is determined by the number of elements in the capacitor stack and the design of the individual elements. A significant improvement in the discharge
inception voltage is obtained by folding the edges of the
electrode foils to produce a more rounded profile.

9.4.3 Bushings

      Bushings are used at the interface between various pow-
er distribution equipments. The high voltage versions contain fluid or resin impregnated insulation which is operated
at a fairly high stress. The construction is typically a
porcelain housing containing a paper element which is wound
and lapped around a central conductor. If the bushing is
integral with other equipment a common impregnant is used,
otherwise a good quality mineral oil is suitable. A well
known design is the capacitor bushing which contains concentric layers of paper dielectric interlayered with aluminium
foils where each foil forms an equipotential screen. The
bushing geometry is well defined and the designer is therefore able to control the stress distribution in both the
radial and axial direction.
      In service, bushings often experience the full extent
of fast rising transients and it is normal for the bushing
insulation to be rated in terms of its ability to withstand
impulse breakdown. The insulation must be carefully dried
during manufacture and with large bushings this may take an
appreciable length of time. Coarse textured papers are
often used to facilitate drying and impregnation.

9.4.4 Transformers

During the past 25 years many new types of synthetic varnish and solid insulation have been used in transformers, nevertheless the mechanical stability and load bearing properties of cellulose are still highly valued by design engineers. In high voltage applications the principal insulation system is fluid impregnated cellulose and in addition to paper tapes, considerable quantities of pressboard and synthetic resin bonded paper are used.

In transformers the average dielectric stress is usually low and it is typically less than one fifth of the stress used in impregnated paper cables and capacitors. One reason for this is the need to maintain a high impulse strength over rather complex insulation geometry. Various mathematical and field plotting techniques are used to determine the field distribution in transformers under both impulse and steady state conditions.

The use of polychlorinated aromatic fluids in transformers has been largely discontinued and many existing units have been retrofilled with alternative fluids. For applications where fire resistance is required there are available organic esters and silicone fluids with a fire point exceeding 300°C. In addition to providing electrical insulation the transformer fluid has an important function as a heat transfer medium. The majority of power transformers rated up to 5 MVA use natural oil circulation those above 25 kvar 3-phase need special fins or tubes to assist the cooling process.

Thermal degradation of the fluid and paper has to be minimised and the operating temperature for the fluid is usually restricted to an upper limit of between 30 and 60°C. The fluids which are chosen for use in large power transformers generally have a high oxidation resistance and may contain inhibitors.

For traction equipment and similar applications it may be advantageous to build transformers with a high operating temperature. In these cases the insulation may be based on high temperature materials such as acetylated paper or aromatic polyamide fibre paper with fluorocarbon fluids. The additional cost of these materials and the increase in onload losses make this technology less attractive for use in power systems.

REFERENCES

9.1   Parkman, N., 1978, 'Some Properties of Solid-Liquid Composite Dielectric Systems'. IEEE Trans. Electr. Insul. Vol.El-13, No.4.

9.2   Parkman, N., and Staight, J H., 1982, 'Polymer-Liquid Interactions in Polypropylene High Voltage Capacitors' BEAMA Int. Electr. Insul. Conf. Brighton.

9.3   Fendley, J. J., and Parkman, N., 1982, 'Effect of
      Impregnation, Compression and Temperature on Electric
      Strength of Polythene and Polypropylene', IEE Proc.,
      Vol.129, Pt.A, No.2.

9.4   Masunaga, M., et al 'Interaction Between Diarylalkane
      and Polypropylene Films in Capacitors, New Materials for
      Electrical Engineering', 1977, Moscow.

9.5   Garton, C.G., 'Dielectric Loss in Thin Films of Insula-
      ting Liquids'. 1941, J.IEE, 88, Pt.II, 103-20.

9.6   Krasucki, Z., H.F.Church and Garton, C G., 'A New
      Explanation of Gas Evolution in Electrically Stressed
      Oil-Impregnated Paper Insulation', 1960, J.Electrochem.
      Soc. 107, No.7, p.598.

9.7   Edwards, D.R., and Melville, D.R.G., 'An Assessment of
      the Potential of EHV Polypropylene/Paper Laminate
      Insulated Self-Contained Oil-Filled Cables'. 1974.
      Underground Transmission and Distribution Conference.

9.8   Devins, J.C. and Reed, C W., 'Crazing of Glassy Poly-
      mers in Oil-Impregnated Cables'. 1970. Dielectric
      Materials Measurement and Applications Conference.

Chapter 10

Applications of solid/gas composites

R. C. Blatcher

The use of solid and gaseous insulating materials in combination falls into two main categories a) in a closed environment such as SF_6 insulated metal enclosed equipment and b) outdoor insulators subject to atmospheric pollution.

10.1 SUPPORT INSULATORS IN GAS INSULATED SUBSTATIONS

10.1.1 Introduction

The use of SF_6 as an insulating and interrupting medium has permitted the delevopment of a compact high voltage metal enclosed range of distribution and transmission switchgear. The system operating voltages extend up to 800 kV. The first SF_6 circuit breakers installed in the U.K. in 1966 at a system voltage of 145 kV were of the dead tank design with SF_6/Air bushings Fig. 10.1. Metal enclosed substations up to 420 kV have been commissioned in the U.K. since 1975. A typical 420 kV substation layout is shown in Fig. 10.2.

The gas operating pressures range from 0.3 MPa in the busbar system to 0.85 MPa in the circuit breaker.

The conductors and interrupters employed in the equipment have to be supported and operated by solid insulating material.

Fig. 10.1   The first SF_6 installation in Great Britain, GEC 132 kV circuit breakers at C.E.G.B. Hall Green Substation.

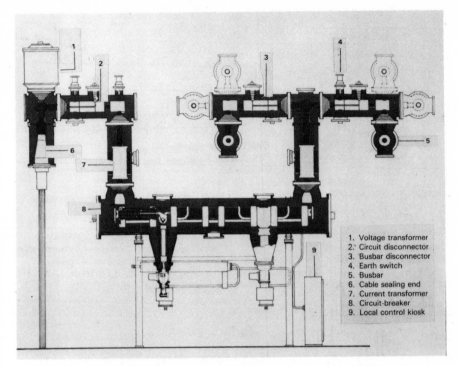

Fig. 10.2 Typical feeder bay for double busbar gas insulated switchgear.

1. Voltage transformer
2. Circuit disconnector
3. Busbar disconnector
4. Earth switch
5. Busbar
6. Cable sealing end
7. Current transformer
8. Circuit-breaker
9. Local control kiosk

10.1.2 Insulator Shapes

The insulators used in the equipment are required for a variety of functions.

Conical insulators Fig. 10.3 are used to support the conductor and to separate gas zones of equal or different pressures. They are also the type generally used to support the conductor in single phase enclosures.

Fig. 10.3 Cast resin cone insulator for 420 kV switchgear.

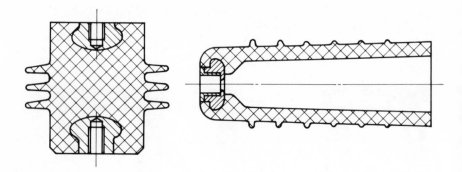

Fig. 10.4   Cast resin support   Fig. 10.5   Cast resin operating
            insulator.   rod.

Solid support insulators Fig. 10.4 are generally used as conductor
supports in three phase enclosures. Circuit breaker interrupter modules
are usually supported on tubular insulators. The mechanism operating
rods Fig. 10.5 can be either tubular or solid. Sheds or ribs can be
added to mitigate the effect of particle contamination.

10.1.3 Electrical Design

10.1.3.1 Solid/gas interface. The effect of introducing a solid
insulating material into a gaseous insulation is to distort the
electrical field due to the different permittivities of the two media.
A measure of the influence of the solid insulation on the inherent
electrical performance of the gas gap is the spacer efficiency i.e. the
ratio of the breakdown strength of the insulator/gas interface with that
of the gas gap without the insulator.

Cookson (10.1) in a review of spacer efficiencies showed that this
value could vary between 0.5 and 0.9. A large part of the variation
used could be attributed to bad contact between the insulator and the
electrode. Various shapes were used but it is not clear from the review
which gives the best profile.

In order to obtain optimum performance of the insulator it is very
necessary to determine the electric field distribution over the surface
of the insulator and to ensure good contact and adequate shielding at
the junction between the metal electrodes and the insulator surface.

Methods of analysing the field distribution are given in Chapter 6.
Similar methods have been used by a number of workers and which can be
illustrated by Takume (10.2) for optimising disc type insulators and
Maskitian (10.3) for post type insulators.

Cook and Trump (10.4) show the effect on breakdown strength of post
type insulators by changing the profile of the insulator and the
electrode moulded into the insulator. The SF$_6$ gas pressure in the
chamber was 0.44 MPa. Under impulse voltages the maximum gas gap
breakdown stress was 25 MVp/m or 57 MVp/m/MPa compared with the
theoretical value of 88 MVp/m/MPa. Providing the stress on the
insulator surface was held below this value of 25 MVp/m or 57 MVp/m/MPa
the voltage breakdown occurred in the gas and not over the surface of
the insulator. A similar value of 50 MVp/m/MPa at 0.3 MPa has also been
confirmed by Nita et al (10.5) and others.

It also has been shown that the relative breakdown voltage over a solid/gas interface is not proportional to gas pressure (10.5, 10.1).

This experimental evaluation was made under clean conditions and consideration has to be given to the reduction in electric strength which will result from surface imperfection, particle contamination and other defects.

10.1.3.2 Solid material. The electric field strength on the surface and within the solid material has to be considered together with its long term ageing properties.

Filled epoxy resin is the most commonly used material, and has been evaluated by Dakin and Studmiarz (10.6) among others. They have shown that a bisphenol A epoxy resin filled with quartz and with no apparent defects has a life in excess of 10^5 hours (11 years) at a stress of 11 MVp/m. This is some 2 to 3 times the normal working stress in current equipment. It had also shown that the material with no apparent defects will withstand limited high voltage tests of short duration without detriment to its long term life. However it is considered desirable to limit the number of over voltage tests applied to any solid insulation.

10.1.4 Effect of Contamination

10.1.4.1 Particle contamination. During the manufacture of SF_6 equipment every care is taken to ensure the exclusion of dust, conducting particles and fibres inside the equipment. However it is not practical to assume that all such particles have been removed. It must also be accepted that the operation of a high speed mechanism in the gas zone will generate small particles (10.7).

The breakdown voltage of SF_6 has been shown to be reduced by particle contamination even without solid insulation (10.8, 10.9). The effect however is small on gas gaps with particles less than 1 mm (10.5, 10.10, 10.11) with a gas pressure of 0.3 Pa.

Insulation surfaces are however more susceptible to particle initiated breakdown. (10.5, 10.10).

It is suggested that dust particles up to 0.5 mm and metallic dust up to 0.05 mm can be tolerated (10.12). The insulator design however will have a marked effect on the reduction in electric strength caused by contamination and the amount of contamination that can be tolerated (10.5, 10.13).

It has been found that contamination of the insulator surface with particles will have a more marked effect, on the A.C. breakdown value (some 30%) than on the impulse breakdown voltage (10.5). Contamination of the insulator surface also results in a lower positive polarity impulse breakdown. This is unlike the characteristics of a gas gap or a clean insulator where the negative impulse breakdown voltage is usually some 10% lower than the positive impulse breakdown value. (10.14, 10.15).

The particles have less influence on the breakdown value at low gas pressure than at higher pressures. The particle size also has a considerable effect on the breakdown levels. Insulating fibres have a small effect but when contaminated with moisture, metal or graphite dust will behave as conducting particles of similar lengths.

It has also been shown (10.5, 10.16) that with conical insulators the junction of the surface of the cone closest to the conductor or enclosure is more susceptible to breakdown (Fig. 10.6).

Particles can be controlled to some extent by the use of so called particle traps (10.17) which are designed to created areas of low or zero stress. The designs appear to be suitable for high pressure SF$_6$ insulated transmission lines but are not considered necessary for substation equipment.

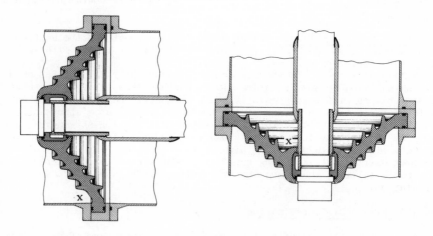

Fig. 10.6 Position of debris in relation to insulator.

10.1.4.2 Contamination by arc decomposition products. The gas in SF$_6$ substation equipment can normally be expected to remain clean and dry throughout the life of the equipment. However circuit breaker gas systems can be contaminated by decomposition products of SF$_6$ during short circuit current arc interruption.

The arc products have variously been reported (10.18, 10.19, 10.20, 10.21) as SF$_4$, S$_2$F$_2$, F$_2$ and when combined with water produced SOF$_2$, SO$_2$F$_2$ and SO$_2$F$_4$.

In addition to the gaseous products very fine metallic fluoride powders are also generated when the arc occurs.

The amount of these decomposition products is directly proportional to the arc energy involved.

The metallic fluoride powders produced during arcing will deposit on surfaces in the enclosures including the insulation. Tests have shown (10.23) that providing the moisture content of the gas is 44 mg H$_2$O/kg SF$_6$ or less no reduction in the breakdown voltage occurred.

Tests on various materials have shown however that with highly acidic decomposition products some materials are more effected than others if the moisture content exceeds 74 mg H$_2$O/kg SF$_6$.

This is particularly the case when comparing silica and alumina filled epoxy resin materials. It has been shown (10.5) that the leakage current can increase by over 4 orders in a few minutes when silica as opposed to alumina is used as the filler.

Under adverse conditions with high moisture contents and highly acidic gas even carefully selected materials can be made to track. These factors emphasis the need to ensure that the moisture content of the gas is controlled.

The use of suitable adsorbent material in the equipment will control both the acidity and the moisture content of the gas and ensure that acceptable levels are maintained.

10.1.4.3 Contamination by water. It has been shown (10.7, 10.22) that
if the moisture content of the gas is high enough to permit water to
condense on the surface of the insulator to form droplets the breakdown
voltage of the insulator will be reduced. However if the dewpoint in
the gas is sufficiently low for the water vapour to form frost on the
insulator by sublimation, no reduction can be expected. However this
also implies that the frost evaporates directly into water vapour and
does not melt into water droplets when the temperature rises.

Power frequency voltage tests (10.7) on insulators in a gas with a
dewpoint greater than $0^{o}C$ and a 50% relative humidity showed a reduction
in electric strength of between 5-17%. However, when using gas with a
dewpoint below $-2^{o}C$ there was little change in electric strength up to
saturation levels.

Current practice in the U.K. is to recommend that the dewpoint of the
gas does not exceed $-20^{o}C$ at working pressure. At this level the
relative humidity at 60C is 16%.

10.1.5 Insulator Material

A variety of materials are used in the manufacture of insulators for
use in SF_6 insulated equipment. The materials are required to have good
electrical and mechanical properties in the working temperature range of
the equipment, to have a low initial moisture content and to be readily
manufactured in the required shape.

The most commonly used material for conductor support insulators is
filled epoxy resin. The system generally used is a bisphenol A epoxy
resin filled with an alumina or dolomite type filler and these have
given many years of satisfactory service. Cycloaliphatic epoxy resins
with similar fillers have also been used but these can have a higher
thermal coefficient of expansion which has to be considered. Chenoweth
et al (10.24) have developed a high resistance material using a
hydantoin epoxy resin filled with fused silica and hydrated alumina.
This has been evaluated by Wooton et al (10.25) for use in SF_6 insulated
high voltage transmission lines where there is a requirement for the
insulator to be undamaged following a breakdown or flashover during a
high voltage test. The energy in the flashover from the capacitance in
the system is high and would normally damage the insulator. This new
system will withstand a number of flashovers involving 18 J/mm of
insulator surface without damage.

Tubular support insulators and operating rods are manufactured from
filled epoxy resin systems or impregnated fabric materials.

The fabrics are generally woven from polyester, glass or in some
cases Kevlar. The glass and Kevlar fabrics will give a component with
a higher tensile modulus than with the polyester fabric. However care
has to be taken to protect the glass fabric from possible attack by the
acidic by-products.

10.2 OUTDOOR INSULATORS

10.2.1 Introduction

Insulators made of ceramic and glass materials have been used in an
outdoor environment since the beginning of the electrical industry. In
the last decade or so attempts have also been made to use organic
materials in the form of cycoaliphatic epoxy resins, silicone rubbers
and other materials.

The insulator electrical design has to have sufficient length to prevent an air breakdown at the appropriate voltage and have sufficient creepage to withstand the degree of pollution apertaining to the site at which they will be located in service.

However a basic problem still exsists in determining the optimum insulator profile and creepage length for a given site condition.

10.2.2 Mechanism of Flashover

The mechanism of flashover of polluted insulators by the formation of dry bands was first reported by Forrest (10.26). Further investigation into the mechanism have been made by Hampton (10.27) and Woodson and McElroy (10.28).

These investigations show that wetting of the pollutant will lead to an increase in the leakage current over the surface of the insulator. Variation in current density caused by the geometry of the insulator and uneven wetting will lead to the formation of dry bands or areas of high resistance. The dry bands will reduce the leakage current and will have to support the voltage applied across the insulator.

It has been shown (10.27) that the voltage stress across the dry band stabilises just below that level required to initiate a discharge in air. If the equilibrium is disturbed then a discharge will occur across the dry band with the current surge limited only by the resistance of the remaining polluted surfaces.

The propagation of the discharges across the dry bands will depend on a number of factors such as the uniformity of the surface contamination, the resistivity of the deposit and the value of the leakage current.

Verma et al (10.29) evaluated three insulators with approximately the same creepage length of 1900 mm but with axial lengths varying from 730 mm to 2200 mm to determine the critical current at which flashover could occur. The insulators were tested in accordance with IEC 507 (10.30) using both the salt fog method and a coating with a solid pollutant, the kieselguhr method.

The critical current I max. was measured as the maximum leakage current in the half cycle preceeding the flashover of the insulator

$$I_c = I_m = \left(\frac{800 \quad Xc}{Vc} \right)^2 \tag{10.1}$$

where :-   800 is a constant
   Xc = Critical arc length in cm = 2/3 creepage length
   Vc = Flashover voltage in Volts

It has been shown (10.31) that for a given type of insulator profile andinsulator will have the same pollution withstand value if the ratio between the applied voltage and the leakage path is the same. Similarly a smaller diameter cylindrical insulator with an identical shed profile will have a pollution withstand value at least equal to that of an insulator with a larger diameter and the same creepage length.

10.2.3 Pollution Severity and Creepage Length

The pollution on an insulator surface can range from light or heavy saline deposits depending on the distance from the coast and the prevailing winds, chemical deposits from industrial complexes and domestic heating systems to sand, dust and saline deposits in desert areas.

The pollution can be activated by light rain, fog or heavy dews. The latter can cause more serious problems with regards to insulator performance than heavy rain.

Tables have been established describing various conditions and classifying different degrees of pollution.

These range from light, medium heavy and very heavy (10.31, 10.32).

The pollution severity for a site can be evaluated from such tables, the behaviour of insulators in a substation or overhead lines in the area or by measurement of the pollution at the intended site.

Five such methods are detailed in (10.33) and are as follows :-

(i)   Equivalent Na Cl deposit density (ESDD). This is expressed in mg Na Cl per cm^2 and enables an equivalent to the amount of the active component of the contaminant to be measured.

(ii)   Surface conductivity. The surface conductance is defined as the ratio of the power frequency current flowing over a sample insulator to the applied voltage. The voltage, about 30 kV·per metre of insulator length, is applied only for a few cycles.

(iii) Surge count. The number of leakage current pulses above a certain amplitude are measured over the insulator at working voltage.

(iv)   I highest. The highest peak leakage current is recorded during a given period over an insulator energised at the working voltage.

(v)   Flashover stress. The insulator flashover stress is the power frequency flashover voltage divided by the overall insulator length. This test involves a number of insulators of different lengths.

These methods can be costly and will of necessity have to cover a reasonable time period.

A large number of tests have been made using these methods to evaluate site conditions and are well documented (10.30, 10.34, 10.35, 10.36, 10.37).

Some areas define the pollution severity by P numbers and use a formula similar to the following to determine the insulator profile

$$P = \frac{0.015 \ La^2 \ x \ Lse}{C \ x \ Um^3} \qquad\qquad (10.2)$$

when :- Um = System highest voltage (line to line)
         La = Bushing flashover length
         Ls = Bushing total creepage length
         Lsp = Bushing protected creepage distance
         a   = Average shed spacing
         b   = Average shed length
         A   = 1.5 - b/2a = Shape factor
         B   = 1.0 - Lsp/2 Ls = Shape factor
         C   = 1.2 = system voltage factor
         Lse = A x B x Ls

These formulae are similar to that proposed by Glasson (10.38).

From these results the required creepage length in kV/mm of system voltage can be derived for a particular shed profile. These range from 16 mm/kV (phase to phase voltage) to 31 mm/kV depending on the pollution severity.

In extreme cases of very severe pollution values of up to 40 mm/kV or 50 mm/kV (10.32) can be used.

The concept of specifying a particular creepage length implies a linearity between withstand voltages under specified pollution conditions and creepage distance. Correction factors may have to be applied to take into account different shed profiles and the diameter of the insulator.

A guide to the relationship between specific creepage distance and various artificial pollution tests is given in (10.39) and may be used as a guide towards initial assessment and selection. However the insulator profile as well as creepage distance will influence the performance. It is therefore necessary to evaluate different insulators before a final selection can be made.

10.2.4 Insulator Profiles

An insulator profile or shed shape can vary in a number of ways and a large number have been evaluated (10.39) to establish withstand salinity values. The shapes range from an antifog type with deep ribs on the underside of the shed to a plain shed with no ribs. Typical insulator shed shapes are shown in Fig. 10.7

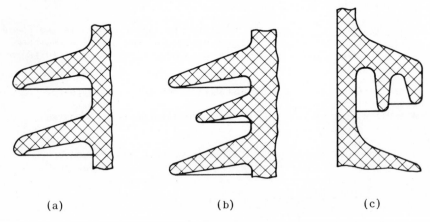

(a)   (b)   (c)

Fig. 10.7 Typical ceramic insulator shed shapes, (a) Plain shed,
(b) Alternate short/long shed, (c) Antifog shed.

Further tests by Ely et al (10.40) also evaluated different shed shapes and showed a good relationship between the salt fog tests and natural pollution at a coastal site.

An evaluation of the self cleaning properties of an insulator using an aerodynamic profile under desert conditions has been made by El-Arabety (10.37) and shows the advantage of using insulators with plain horizontal sheds.

The spacing of the sheds, the overhang and the ratio of creepage

length to flashover length have to be considered together with the
mechanical strength of the sheds in order to produce an insulator which
will perform satisfactorily in service. Various I.E.C. publications
make recommendations on these latter points.

10.2.5 Coating of Insulators

In order to reduce the effect of pollution on the flashover voltage
of insulators, coatings of various greases have been applied to the
surface of the insulator. (10.41, 10.42). The greases which can be
either of hydrocarbon or silicone materials envelop the pollution and
thus prevent discharges occuring.

Hydrocarbon greases soften on heating, melt at the site of any
discharge and envelop the pollution. The tendency to soften on heating
restricts their use to temperate climates. The grease is applied about
3 mm thick either by hand or by dipping. The removal of the grease at
the end of its useful life can be rather time consuming.

Silicone greases because of their temperature stability can be used
at higher temperatures. They are usually applied at least 1 mm thick.

Silicone greases are more costly than hydrocarbon greases.

It has been shown that if too thin a layer of grease is applied glaze
damage can result.

A coating of silicone elastomer has also been applied to the surface
of insulators in thickness of about 1 mm. Tests made under desert
conditions (10.37) showed that an insulator coated with silicone rubber
withstood exposure to severe weather conditions for 3 years without
damage and also had the lowest average pollution density of the
insulators tested, showing a good self cleaning characteristic.

Tests also made by Hall and Orbeck (10.43) showed that with correctly
selected silicone elastomers the coating will show an improvement in the
pollution flashover value compared with uncoated insulators.

Silicone elastomer material would appear to be an improvement over
greasing in cost and can be more readily renewed than grease coatings.

10.2.6 Cleaning of Insulators

Insulators can either be cleaned manually or by the use of water jets
to remove pollution deposits.

Washing techniques can be carried out with the insulators energised
(10.44, 10.45, 10.46) using a fixed spray installation or a manually
controlled portable jet.

The frequency of washing will depend on the site condition but should
be adequate to maintain the insulator in a clean condition. Monitors
can be used to measure the pollution and types are available to give
automatic control of the washing frequency.

10.2.7 Insulator Materials

The most commonly used materials for outdoor insulators are glass and
ceramics. The properties of these are well documented and embodied in
various national and international specifications.

The use of organic materials such as cycloaliphatic epoxy resins,
silicone rubbers and polyolefines is growing in some areas but their
general adoption has to be treated with care. The problems encountered
in their use for overhead lines is well documented by Cojan et al
(10.47). The testing and evaluation of these materials is detailed in
Chapter 15.

Insulators manufactured from a filled cycloaliphatic resin system however have been in use outdoors in the Middle East for several years on systems with voltages ranging from 33 kV to 115 kV.

Acknowledgement

The author wishes to thank his colleagues for their assistance and the Directors of G.E.C. High Voltage Switchgear for permission to publish.

REFERENCES

10.1   Cookson, A.H., "Electrical breakdown for uniform field in compressed gases", 1970, Proc. IEE, 117, 269-280.

10.2   Takuma, T., and Watanabe, T., "Optimal profiles of disc-type spacers for gas insulation", 1975, Proc. IEE, 122, 183-188.

10.3   Mashikian, M.S., Whitney, B.F., and Freeman, J.J., "Optimal design and laboratory testing of post-type spacers for three phase SF6 insulated cables", 1978, IEEE Trans., PAS-97, 914-918.

10.4   Cooke, C.M., and Trump, J.G., "Post-type support spacers for compressed gas-insulated cables", 1973, IEEE Trans., P.E.S. Winter Meeting Paper, T73, 121-1.

10.5   Nitta. T., et al., "Factors controlling surface flashover in SF6 gas insulated systems", 1978, IEEE Trans., PAS-97, 959-965.

10.6   Dakin, T.W., and Studniarz, S.A., "The voltage endurance of cast epoxy resins", 1978, International Conference on Electrical Insulation.

10.7   Hogg, P., Schmidt, W., and Strasser, H., "Dimensioning of SF6 metalclad switchgear to ensure high reliability", 1972, CIGRE, Paper 23-10.

10.8   Kuwahara, H., et al., "Effect of solid impurities on breakdown in compressed SF6 gas", 1974, IEEE Trans., PAS-93, 1546-1555.

10.9   Diessner, A., and Trump, J.G., "Free conducting particles in a coaxial compressed-gas-insulated system", 1970, IEEE Trans., PAS-89, 1970-1978.

10.10   Johnson, B.L., Doepken, H.C., and Trump, J.G., "Operating parameters of compressed gas insulated transmission lines", 1969, IEEE Trans., PAS-88, 369-375.

10.11   Cookson A.H., et al., "Recent research in the United States on the effect of particle contamination reducing the breakdown voltage in compressed gas-insulated systems", 1976, CIGRE, Paper 15-09.

10.12   Roth, A.W., and Owens J.B., "Dielectric investigations of components for gas insulated switching stations and conductor systems for extra high voltage", 1970, CIGRE, Paper 23-03.

10.13   Itaka, K., and Ikeda, G., "Dielectric characteristics of
        compressed gas insulated cables", 1970, IEEE Trans., PAS–89,
        1986–1993

10.14   Hampton, B.F., and Fleming, S.P., "Impulse flashover of particle
        contaminated spacers in compressed SF_6", 1973, Proc. IEE, 120,
        514–518.

10.15   Morii, K., et al., "Development of 500 kV gas insulated
        switchgear and its applications", 1973, IEEE Trans., P.E.S.
        Winter Meeting, Paper T73, 033–8.

10.16   Kuwahara, H., et al., "Effect of solid impurities on breakdown in
        compressed SF6 gas", 1974, IEEE Trans., P.E.S. Winter Meeting,
        Paper T74, 189–7.

10.17   Dale, S.J., and Hopkins, M.D., "Method of particle control in SF_6
        insulated CGIT systems", 1982, IEE Trans. PAS 101, 1654–1663.

10.18   Edelson, C.A., et al., "Electrical de composition of sulphur
        hexafluoride", 1953, Ind. and Eng. Chem., 45, 2094–2096.

10.19   Manion, J.P., et al., "Arc stability of electronegative gases",
        1967, IEEE Trans., E1–2, 1–10.

10.20   Miyamoto, T., and Kamatani, A., "Application of a time resolved
        type panoramic mass spectometer to the studies of arc quenching
        reactions in SF_6 gas", 1965, J. IEE., Japan, 85, 675–684.

10.21   Hirooka, M., et al., "Chemical properties of sulphur hexafluoride
        gas", 1970, Mitsubishi Denki Giho, 44, 1175–1182.

10.22   Tsunero, U., et al., "Practical problems on SF6 gas circuit
        breakers", 1971, IEEE Paper, 71 TP 64–PWR.

10.23   Harris, A.F., "Insulating surfaces associated with high voltage
        metalclad SF6 switchgear", 1975, IEE Digest No. 1975/26, 1/1 - 1/6

10.24   Chenoweth, T.E., et al., "Epoxy casting composition for UHV
        service in compressed SF6 insulated equipment", 1977,
        13th Electrical/Electronics Insulation Conference, 322–326.

10.25   Wootton, R.E., and Emery, F.T., "A test for the effect of high
        energy arcs on the flashover strength of insulators in compressed
        SF6", 1980, IEEE Trans., E1–15, 74–80.

10.26   Forrest, J.S., "The characteristics and performance in service of
        high voltage porcelain insulators", 1942, J. IEE., 89, Pt. 11,
        60–92.

10.27   Hampton, B.F., "Flashover mechanism of polluted insulation",
        Proc. IEE., 111, 985–990.

10.28   Woodson, H.H., and McElroy. A.J., "Insulators with contaminated
        surfaces, Part 111 : Modeling of dry zone formation", 1970,
        IEEE Trans., PAS–89, 1868–1876.

10.29 Verma, M.P., et al., "The criterion for pollution flashover and its application to insulation dimensioning and control", 1978, CIGRE, Paper 33-09.

10.30 IEC 507, 1975, "Artificial pollution tests on high voltage extra insulators to be used on A.C. systems".

10.31 CIGRE Working Group 04, "A critical comparison of artificial pollution test methods for H.V. insulators", 1979, Electra, 64, 117-136.

10.32 Australian Standard No. 1265, 1974, "Bushings for alternating voltages above 1000V".

10.33 CIGRE Working Group 04, "The measurement of site pollution severity and its application to insulator dimensioning for A.C. systems", 1979, Electra, 64, 101-116.

10.34 Houlgate, R.J., et al., "The performance of insulators at extta and ultra high voltage in a coastal environment", 1982, CIGRE, Paper 33-01.

10.35 He Pei-Zhong, and Xu Cheng-Dong, "The tests and investigation results on naturally polluted insulators and their application to insulation design of power systems in the polluted areas", 1982, CIGRE, Paper 33-07.

10.36 El-Koshairy, M.A.B., et al., "Pollution performance of high voltage insulator strings in a desert environment", 1982, CIGRE, Paper 33-09

10.37 El-Arabaty, A., et al., "Selection of insulators suitable for operation in contaminated environments with reference to desert conditions", 1980, CIGRE, Paper 33-11.

10.38 Glasson, G.T., "The flashover of polluted insulation at power-frequency voltages", 1969, Pro. Inst. of Eng. Australia, Paper No. RVE/E5763.

10.39 Lambeth, P.J., et al., "International research on polluted insulators", 1970, CIGRE, Paper 33-02.

10.40 Ely, C.H.A., et al., "Artificial and natural pollution tests on outdoor 400 kV substation insulators", 1971, Proc. IEE, 118, 99-109.

10.41 Lambeth, P.J., "Surface coating for H.V. insulators in polluted areas", 1966, Proc. IEE, 113, 861-869.

10.42 Davydova, L.I., et al., "On optimum thickness of water-repellent covering of high voltage insulation", 1980, Sov. Power Eng., 9, 724-728.

10.43 Hall, J., and Orbeck, T., "Evaluation of a new protective coating for porcelain insulators", 1982, IEEE Trans., PAS-101, 4689-4696.

10.44   El Sayed A.H. Aly., "Results of tests on dielectric strength of low pressure water jet for live washing of the 500 kV transmission line in the Arab Republic of Egypt", 1974, <u>CIGRE</u>, Paper 33-11.

10.45   Last, F.H., et al., "Live washing of h.v. insulators in polluted areas", 1966, <u>Proc. IEE</u>, <u>113</u>, 847-860.

10.46   Cakebread, R.J., et al., "An automatic insulator washing system to prevent insulator flashover due to pollution", 1977, IEE Conference", The design and application of EHV substations", 35-39.

10.47   Cojan, M., et al., "Polymeric transmission insulators : their application in France, Italy and the U.K.", 1980, <u>CIGRE</u>, Paper 22-10.

Chapter 11

Evaluation of liquid insulants

A. Kallinikos

11.1 INTRODUCTION

Insulating liquids are used in electrical equipment
such as transformers, switchgear, cables, bushings,
capacitors, mainly as insulants but also in some cases
(transformers and cables) as cooling media. In certain
cases these liquids act as arc extinguishers (switchgear)
and even as lubricants where moving parts are present
(switchgear, circulating pumps and tap changing equipment).

As shown in chapter 7 different types of equipment
require dielectric fluids with special properties.
Transformers and most other types of electrical equipment
need liquids with high electrical strength, impulse strength
and resistivity but low dielectric dissipation factor (loss
tangent). The dielectric must also possess high specific
heat and thermal conductivity along with low viscosity and
pour point in order to keep the equipment cool. Good
chemical and thermal stability and gas absorption properties
are also desired.

Capacitors also require dielectric fluids with high
discharge resistance and switchgear needs fluids with arc
quenching properties.

Other desirable properties are high flash points, if
possible liquids should be non-flammable or at least have
high fire points and for ease of handling and for ecologi-
cal purposes they should be non-toxic.

For the evaluation of dielectric liquids the determination of the electric properties is obviously of prime
importance. It is also important to determine not only
whether the fluid possesses good electrical properties but
also whether these properties can be maintained during the
life of the equipment with or without processing or small
additions to the liquid of materials (e.g. oxidation
inhibitors, passivators, acid scavengers) to help the
dielectric fluid to maintain as nearly as possible its
original electrical properties.

The chemical stability of the fluid can be assessed by
the determination of its resistance to oxidation. It is
also important to determine the water content and acidity
of the fluid as these two can greatly affect its electrical
properties. Other properties that must be determined are
the gas absorption, flammability, toxicity, and whether

breakdown products produced by arcing or overheating are present. Dissolved gas-in-oil analysis can determine whether a transformer or bushing are free from electrical faults.

The following describes the test methods now employed to determine or control these properties.

11.2 ELECTRIC STRENGTH OF LIQUIDS

11.2.1 Breakdown Testing

The electric strength of an oil is the breakdown voltage of the oil between plane, spherical or hemispherical electrodes set a certain distance apart (2.5mm) and located in a test cell. The following procedure is used according to BS148:1972 for the determination of the electric strength of an oil using spherical electrodes (11.1).

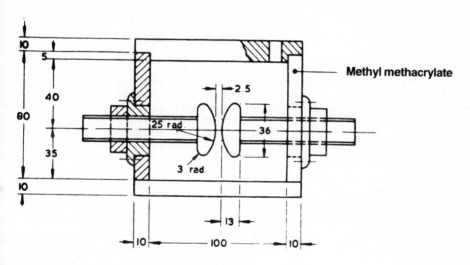

Fig.11.1   Typical Breakdown Test Cell

The test cell, (Fig. 11.1) is cleaned by rinsing with the test oil several times. The vessel containing the test oil is gently agitated to ensure a homogeneous distribution of impurities contained in the oil without causing undue formation of air bubbles. Immediately after, the oil sample is poured slowly into the test cell, in a dry and dust-free area. The oil temperature at the time of the test shall be the same as that of the ambient air preferably in the neighbourhood of $20^{o}C$ (15° - $25^{o}C$).

Once the oil appears free from air bubbles an increasing voltage of frequency 40Hz to 52Hz is applied to the electrodes, starting from zero and increasing at the rate

of 2kV/s upto the value which produces breakdown.   The
circuit is opened manually if a transient spark (audible
or visible) occurs between the electrodes or automatically
if an established arc occurs.   In the latter case the
automatic switch shall break the circuit within 0.02s.
The test is carried out six times on the same cell filling.
After each breakdown, the oil is gently stirred between the
electrodes by means of a clean, dry glass rod, to disperse
the carbon produced by the action of the arc in the oil,
and allow to stand for 5 minutes to allow the air bubbles
to escape before a new breakdown test is started.
      The breakdown voltage is the voltage reached during
the test at the time the first spark occurs between the
electrodes, whether it be transient or established.   The
electric strength is the arithmetic mean of the six results
obtained.   The reporting of the result shows all six
breakdown voltages obtained and the average of these results.
      Alternatively the one minute withstand at the stated
test voltage is carried out using a 2.5 mm gap spacing.
(11.2)
      Here only one oil filling need be tested and a pass is
recorded if the test voltage is withstood for 1 minute.
If it fails this test the same cell filling may be subjected to two further tests, and if in both test the oil
withstands the application of test voltage for 1 minute,
the sample is deemed to have passed the test.   Gentle
stirring of the oil must be carried out between tests to
disperse the carbon formed between the electrodes and any
air bubbles formed should be allowed to escape before
retesting.

11.2.2 Meaning of Test Results

      Minimum Test Voltages for the 1 minute withstand test
and 2.5 mm gap are recommended in a Code of practice
BS 5730 (11.2) are as follows for various categories of
equipment :-

           Categories A and D equipment
           170kV and below   - 30kV(40kV for 4mm
                                                        gap)

           All categories other than
           A and D 170kV and below   - 22kV(30kV for 4mm
                                                        gap)

      Whereas the minimum Breakdown Voltages recommended
for the average of six breakdowns rapid rise test using
2.5 mm gap (11.2) are as follows ;-

           Categories A and D above 170kV - 55kV

              "   A and D 170kV &below- 45kV

           All categories other than

                      A and D   - 30kV

Category A, according to BS 5730, is for power transformers and associated equipment which requires to be critically and continuously in service, or any power transformer of any rated voltage which are fitted with a refrigerator breather.

Category D is for voltage transformers and equipment that have rated voltage above 170kV or current transformers that are rated voltage above 40kV (11.2). Further recommendations for other types of equipment can be obtained by reference to BS 5730 (11.2).

11.3 WATER CONTENT OF LIQUIDS

11.3.1 Methods of test

There are two main instruments which are used for the determination of the moisture content of fluids. The first is a direct measure of water content by the well-established Karl Fischer titrimeter method (11.3) and this instrument is described below. The second, which is described later determines the moisture content by measuring the dew point temperature of the fluid. (11.4) (11.5). Dew point temerature is converted (by means of a dual scale) to the corresponding water vapour pressure, and from this the relative humidity follows directly.

In any equilibrium system, the moisture content of all the materials present, whether solid, or gaseous is closely related. Thus the measurement of the moisture level in one material will indicate the moisture content of the others. For example, the water present in transformer paper insulant can be determined from the vapour pressure of water in the oil containing the paper.

11.3.2 The Karl Fischer method

The concentration of water in oil can be determined by the Karl Fischer method. The method uses the reaction of water with Karl Fischer reagent which is a solution of iodine in SO_2 in a pyridine/methanol mixture.

The reactions may be simply expressed as follows:-

$$H_2O + I_2 + SO_2 + 3C5H5N \longrightarrow 2C_5 H_5 NHI + C_5 H_5 NSO_3 - (11.1$$

$$C_5 H_5 NSO_3 + CH_3OH \longrightarrow C_5H_5NH. SO_4CH_3 \qquad -(11.2)$$

The water in oil is directly titrated with Karl Fischer reagent and the end point is detected electrometrically using platinum electrodes.

A mixture of chloroform/methanol 1:1 v/v is introduced into the reaction vessel which contains the platinum electrodes and a magnetic stirrer, and neutralised with Karl Fischer reagent. A known weight of the oil is then added by means of a syringe taking precautions to exclude atmospheric moisture and again neutralised with Karl Fischer reagent with continuous stirring. The strength, or equi-.

valent weight of Karl Fischer reagent is similarly determined by adding a known quantity of water into a previously
neutralised chloroform/methanol mixture and titrating to
end point.

The instrument utilizes this property of the Karl
Fischer reagent by monitoring the precise amount needed
to neutralise traces of moisture in the sample (which is
dissolved or immersed in a solvent mixture) and comparing
this with the amount required to neutralise a known volume
of water. An electrometric circuit isused to detect the
point at which neutralisation has occurred, and details
of this circuit and the method are given in references
(11.3) and (11.5).

The method is similar to BS 2511, 1970 and ASTM D
1533-77 and will replace the method in the IEC Document 10A
(Secretariat) 46.

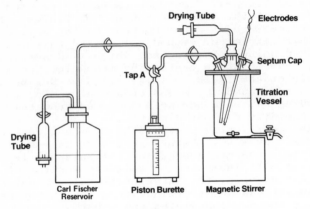

Fig.11.2   Karl Fischer Apparatus

<u>11.3.2.1 Description of Apparatus</u>. The apparatus, (Fig.
11.2) consists of a glass titration vessel, with stirrer,
into which Karl Fischer reagent can be accurately dispensed
from a reservoir by means of a calibrated piston-type
burette. The titration vessel also has inlets for solid
or liquid samples, and is additionally equipped with
electrodes which via an electrometric circuit give an
indication of the end point of titration.

<u>11.3.3 The Relative Humidity Method</u>

The relative humidity of the oil is the water vapour
pressure corresponding to the dew-point temperature divided
by the saturation vapour pressure corresponding to the
test temperature multiplied by 100%.($\frac{VP}{SVP}$ x 100%). A
small volume of previously dried air contained in a spiral
tube made of a semi-permeable membrane is circulated by
means of a small pump so that it absorbs water vapour passing through the membrane from the agitated oil sample until
equilibrium is attained. (11.5)

The dew-point is determined by the formation of dew on
a thermo-electrically cooled mirror. The dew on the mirror

is picked up by a light beam directed on to the mirror surface and reflected onto a photo-transistor. Attenuation of the light beam by dew, changes the current supplied by the amplifier and hence controls the mirror temperature at the dew point of the air. (Fig. 11.4)

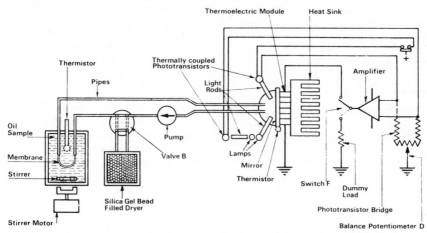

Fig.11.3   Schematic diagram of a Water Vapour Pressure meter

The relative humidity of the oil depends on the solubility of water in oil and is therefore a more useful value to the engineer than the water content of the oil. Oils with higher water solubilities can tolerate higher water contents for the same RH.

11.3.3.1 Principle of Operation of RH Measurement. The principle of operation of the instrument is shown in (Fig. 11.3). A small fixed quantity of air is circulated by an electromagnetic diaphragm pump through a tubular membrane permeable to water vapour, the membrane being totally immersed in the liquid sample. Also included in the air circuit is an electrically cooled mirror on which condensation can take place.

After a period of circulation, and by virtue of the membrane, the water vapour pressure in the circulating air establishes equilibrium with that in the sample. An optoelectronic method is then used to control the mirror temperature at the exact dew point of the air. This is achieved by reflecting a light beam from the mirror onto a photo-transistor so that if a dew point starts to form, the light beam is attenuated and the photo-transistor then regulates the current to the thermoelectric cooler (11.4) (11.5).

The maximum values of RH for power equipment are:-

(a) For low voltage applications upto 33kV. RH upto   50%
(b) For higher voltage   "   33kV-275kV   "   "   20%
(c) For voltages higher than 275kV   "   "   10%

11.3.4 Crackle Test

If a crackling noise is produced when oil is heated in a test tube or other container, the oil is saturated with water (i.e. free droplets of water are present). (11.1)

This is a very crude test and as saturation in insulating oils varies from 70 ppm for a new petroleum oil to 2000 ppm in an Ester Transformer/Capacitor Oil, it is evident that only the extreme cases of contamination can be detected by this method.

11.3.5 Other Methods

Methods of determining water in oils, by reaction with Calcium Carbide and measuring gas evolved by gas chromatography, by electrolytic hygrometer, or by changes in alumina resistor/capacitors are all largely experimental and require considerably more development.

11.4 TESTING FOR TOXICITY AND FLAMMABILITY

11.4.1 Toxicity

Dielectric fluids fall mainly into two groups. Those based on mineral oils and those based on synthetics. The mineral oils are composed of mixtures of aliphatic, naphthenic and aromatic hydrocarbons.

The synthetics are synthetic hydrocarbons e.g. (a) poly-olefins and alkylated aromatic hydrocarbons, (b) halogenated hydrocarbons, e.g. askarels (c) organic esters, which includes esters of vegetable origin e.g. castor oil and (d) silicones and fluorocarbons.

The toxicity of organic compounds is usually determined by exposing animals (rabbits and/or mice) or fish to various concentrations of the chemicals over prolonged periods and studying their effects. This is obviously a time-consuming and very specialised procedure, needing highly qualified personnel.

Toxicity can also be determined indirectly by association with known similar chemicals of known physiological activity and properties. Obviously this is not absolutely satisfactory or safe but it is a useful guide.

11.4.1.1 Mineral Oils. Pure aliphatic mineral oils are considered safe(e.g. medicinal liquid paraffin). Petroleum insulating oils do however contain various types of aromatic hydrocarbons

Of the aromatic hydrocarbons, the alkyl substituted monoaromatics (e.g. toluene, dodecyl benzene) are considered non-carcinogenic. (11.6) Some poly nuclear aromatics are, however, known carcinogens. (11.6) Determination of the content of this type of compounds in mineral oils or other fluid is now considered to be of great importance. This can be carried out by infra-red spectroscopy. A suitable method is described in the CERL Laboratory Report ref.(11.6)

11.4.1.2 <u>Synthetics</u> Of the synthetic oils all aliphatic esters, silicones and aliphatic hydrocarbons are considered non-toxic and should be handled as per ordinary aliphatic mineral oil. Halogenated hydrocarbons whether aliphatic or aromatic are considered toxic. Polychlorinated biphenyls (PCB's) are considered as probably carcinogens.
     All these materials can be successfully characterised by infra-red spectroscopic examination using the method given in ref (11.6).

11.4.1.3 <u>Degradation Products</u> Degradation of dielectric fluids to produce gaseous products is usually the product of fire arcing or overheating under oil. It is important that these products are not highly toxic. The degradation products obtained from hydrocarbons esters and silicones are considered to be non-toxic being hydrocarbons e.g. acetylene, ethylene, ethane and methane, and hydrogen. The only toxic product produced is carbon monoxide, but this is produced in only small quantities.
     Chlorinated hydrocarbons produce chlorine and hydrogen chloride, both toxic gases. Fluorocarbon and Fluorochlorocarbons produce fluorine, chlorine, hydrogen fluoride and hydrogen chloride which are also toxic gases (see chapter 7).
     Toxic compounds of this type can be detected and analysed by colour change tubes (e.g. Draeger Normalair) or by gas chromatography and infrared spectroscopy (11.6).

11.4.2 <u>Flammability</u>

     The Flammability of a liquid can be assessed by the determination of its flash point and fire point.
     The flash point of a liquid is the lowest temperature at which the vapour above it will momentarily ignite in air when a flame is applied to it. The fire point is the lowest temperature of the liquid at which sufficient vapours continue to be formed after ignition to maintain the fire for five seconds.
     The flash point is usually used to decide the highest temperature at which it is safe to use an insulating oil. This can be determined by means of a closed or open cup. Closed cup (Pensky-Martens) to BS4688 or Open Cup (Cleveland to BS4689) (11.9). The difference between closed cup flash point and an open cup fire point is approximately 10°C. (11.8)
     The method in all cases involves the lowering of a small flame into the gas space above a heated volume of liquid and noting when the first flash of vapour igniting occurs, or in the case of fire points when the flash occurs and a flame is maintained.

11.5 <u>GASSING OF INSULATING OILS UNDER THE INFLUENCE OF</u>
      <u>AN ELECTRIC DISCHARGE</u>

     Insulating oils exhibit either gas absorbing or gas evolving properties under the influence of electric discharge at the gas-oil interface. For closed systems with little or no gas phase such as capacitors and cables, hydrogen absorbing oils are required, whereas for free breathing or nitrogen blanketed high voltage equipment, e.g. transformers, air or nitrogen absorbing oils are preferred.

The gassing characteristics of insulating oils are therefore important for the design and manufacture of high voltage equipment and methods for assessing these properties have been developed.

The gassing properties of an oil depend on its composition and especially its aromatic content; oils with aromatic content of 11% are usually gas absorbing and oils with lower aromatic content are gas evolving.

The aromatic content of the oil also governs its oxidation stability, the lower the aromatic content the higher the oxidation stability. Hence, for free breathing electrical equipment, e.g. transformers, a compromise is usually made in order to satisfy both requirements. For closed systems, such as capacitors and cables with little or no gas phase, oxidation stability is of less importance and therefore oils with high aromatic content which are also gas absorbing can be used.

The gassing behaviour of an oil is also affected by the physical environment. Thus an oil can be gas absorbing under an atmosphere of hydrogen or gas evolving under an atmosphere of nitrogen. The same oil under atmosphere of air may show gas absorbing characteristics by the fact that the electrical discharge can convert some of the oxygen and nitrogen to nitrogen oxides thus reducing the volume of the gas above the oil surface.

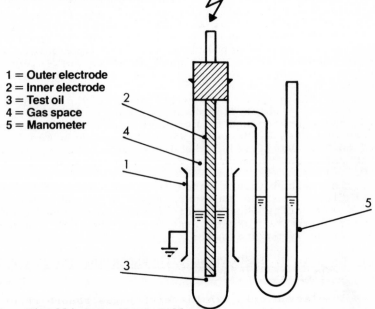

1 = Outer electrode
2 = Inner electrode
3 = Test oil
4 = Gas space
5 = Manometer

Fig.11.4   Pirelli type Test Cell

Pilpel and Reynolds (11.9) using a gassing cell consisting of a glass cell with a central glass electrode (Fig. 11.4) in which the volume of gas above the oil is comparable to the volume of the oil, claim that all hydrocarbon oils, irrespective of their constitution evolve gas under an atmosphere of nitrogen, the amount increasing as the paraffinic content increases at the expense of the aromatic content.

Petersen reviewed the results (11.10) obtained by means of several types of gassing cells. He also carried out a number of experiments using a Pirelli type gassing cell (Fig. 11.4), and a cell in which the volume of oil was much bigger than the volume of gas. It was found in general that the results obtained depended on the design of the test cell, the value of applied voltage, the type of gas, the temperature of the oil and the relative volumes of gas and oil.

A general observation was that oils with high aromatic content were gas absorbing with a hydrogen gas phase but usually gas evolving with a nitrogen gas phase. With an air gas phase oils were usually gas absorbing initially but gradually become gas evolving.

Results obtained by a number of experimenters using a gassing cell where the gas volume above the oil is small compared to the volume of oil, showed gas absorbing properties (CIGRE method as described in IEC - 10A Secretariat 48) (11.11).

The same experimenters using a Pirelli test method according to IEC. Doc.10A (Secr.) 44 and applied voltage 10kV (field strength 3.84 kV/mm) showed that oils with aromatic contents from 4 - 14% inclusive were gas evolving under an atmosphere of N_2 but were gas absorbing under an atmosphere of H_2 where the aromatic content is>10%.

11.5.1   Current Test Methods

The test cells agreed by IEC for carrying out the determination of the gassing characteristics of hydrocarbon oils is the Pirelli test cell or modification of it, the Nederbragt Cell and the Siemens discharge tube.

Summary of the method

The oil is degassed, saturated with the test gas (Nitrogen or Hydrogen) and then subjected to a radial electrical stress under the following experimental conditions:-

Voltage	10kV
Frequency	50 - 60 H
Duration	Upto 24 hrs
Temperature	Upto 80°C
Gap between electrodes	2 mm

11.6 DISSOLVED GAS ANALYSIS.

It has been known for some time that insulating oil
in high voltage equipment can break down under the influence of thermal and electrical stresses to produce low
molecular weight hydrocarbon gases, hydrogen and carbon
oxides. High voltage transformers are often fitted with
Buchholz relays to collect the gases, monitor the rate of
gas formation and give warning when a transformer was gassing by sounding an alarm. Any gas, whether inert or
produced by a fault, once released into the Buchholz could
trip the relay. Gases produced by electrical faults are
usually combustible, where as inert are noncombustible.
The differentiation between combustible and noncombustible
gases was carried out by igniting the gases escaping from
the tap of the Buchholz relay, (11.12). If ignition took
place it could be safely assumed that electrical faults
were present. Failure to ignite the gases was no proof
that combustible gases were absent, but only that their
concentration was not high enough to cause ignition.

Chemical methods were developed to detect the presence
of acetylene, hydrogen and carbon monoxide in Buchholz gas
samples and thus differentiate between atmospheric gases
and gases produced by high temperature degradation of oil
and cellulose insulation, (11.13).

Although these chemical analyses enabled the electrical engineer to decide whether an electrical fault was present and to some degree it's severity, they were obviously
inadequate.

The development of analytical instruments such as the
infrared spectrophotometer, the mass spectrometer and the
gas chromatograph which are capable of characterising and
quantifying very low concentrations of all the gases
produced by electrical faults, has enabled the chemist to
analyse Buchholz gas samples precisely and very often detect
the presence of electrical faults in their early stages.
The correlation of the analysis of Buchholz gas with inspection of the equipment from which the gas was obtained,
enabled the chemist and electrical engineer to assign types
of faults with certain gases, (11.14), (11.15).

Table 11.1 shows the three main types of electrical
faults and the gases produced in oil.

TABLE 11.1 Principal gases from electrical faults

Corona discharge	H_2	CH_2
Heating faults	C_2H_4	H_2
Arcing	C_2H_2	H_2

The gas chromatograph has been established as the
most convenient instrument for the analysis of transformer
fault gases. It possesses adequate sensitivity, is easy
to use once installed and is reliable.

As electrical faults usually take place under the
surface of oil, and gases produced by electrical faults are

soluble in oil, these gases can be removed from the oil by vacuum degassing and then can be analysed by gas chromatography, (11.16) (11.17).

Table 11.2 gives the results obtained by simulating certain faults in the laboratory (11.17).

TABLE 11.2   Gases formed from arcing in various fluids

% by volume

	H_2	CH_4	C_2H_2	C_2H_4	C_2H_6	CO	CO_2	HCl
Transformer oil	60	3.3	25	2.1	0.05	0.1	0.1	-
Silicone Fluid	64	1.2	5.7	0.2	0.06	2.1	ND	-
Phosphate Esters	69	1.18	15.8	0.4	0.03	2.6	ND	-
Trichlorotrifluoroethane	Mixture of CF_4 and CCl type compounds only							
Arclor	48	0.15	22	0.4	0.2	0.3	0.5	20
Perfluorodecalin	-	-	-	Mainly CF4		-	-	-

11.6.1   Solubility of Gases

The solubility of fault gases in oil varies considerably and therefore partial degassing will not give the true concentration of these gases, hence a correction must be applied to the results to take account of these solubilities.

Table 11.3 gives the solubilities of the main fault gases, nitrogen and oxygen in oil. (11.18).

TABLE 11.3   Solubility of Gases in Transformer Oil
At 25°C and 760 mm Hg

Hydrogen	H_2	7%	volume by volume
Methane	CH_4	30%	" " "
Ethylene	C_2H_4	280%	" " "
Ethane	C_2H_6	280%	" " "
Acetylene	C_2H_2	400%	" " "
Carbon Monoxide	CO	9%	" " "
Carbon Dioxide	CO_2	120%	" " "
Oxygen	O_2	16%	" " "
Nitrogen	N_2	8.6%	" " "

In order to avoid the necessity for corrections, total degassing of the oil can be carried out by vacuum stripping the gases several times by means of a Toepler pump or a Torricelli type apparatus. This can achieve the extraction of more than 95% of the fault gases.

11.6.2 Interpretation of results

The analysis is usually carried out for hydrogen, methane, ethylene, ethane, acetylene, carbon monoxide and carbon dioxide. The absolute concentration of fault gases can give an indication of the state of the insulation of the electrical equipment, whereas the relative concentration of these gases can give the type of the fault present. Very often, however there are more than one active faults present which makes interpretation of the results difficult.

Very simply, faults can be classified as partial discharge of low energy which produces mainly hydrogen with a little methane, discharge of high energy (arcing) which produces mainly acetylene and hydrogen with smaller quantities of ethylene and other hydrocarbon gases, and thermal faults which produce hydrogen, ethylene, methane and ethane, the higher the concentration of ethylene relative to ethane the higher the fault temperature.

The presence of carbon monoxide with any of the above three faults in concentration comparable with the rest of the other fault gases could signify that solid insulation e.g. cellulose is involved with the fault.

Carbon dioxide and carbon monoxide are formed by the natural ageing of the oil and cellulose insulation and are usually present in oil in the proportion of approximately 10 to 1. Care therefore must be taken during the interpretation of the results, especially when dealing with transformers which have been in service for many years. Obviously, the older the transformer the higher the concentration of carbon oxides can be expected.

The presence of diacetylene has been reported to be associated with arcing in oil, furane with arcing involving cellulose (11.19) and methanol with arcing, involving phenolic resin insulation (11.15).

11.6.3 Ratio Method of Diagnosis

A more systematic attempt for the qualitative interpretation of the results of gas-in-oil analysis is the Rogers (11.8) Ratio method, which uses the ratios of five gases, and when these fall within certain limits faults are assigned as Table 11.4.

$\dfrac{CH_4}{H_2}$	$\dfrac{C_2H_6}{CH_4}$	$\dfrac{C_2H_4}{C_2H_6}$	$\dfrac{C_2H_2}{C_2H_4}$	Diagnosis
> 0.1 < 1.0	< 1.0	< 1.0	< 0.5	Normal
≤ 0.1	< 1.0	< 1.0	< 0.5	Partial Discharge -Corona
< 0.1	< 1.0	< 1.0	≥ 0.5 ≤ 3.0 or>3.0	Partial discharge-Corona with tracking
> 0.1 < 1.0	< 1.0	≥ 3.0	≥ 3.0	Continuous discharge
> 0.1 < 1.0	< 1.0	≥ 1.0 ≤ 3.0 or >3.0	≥ 0.5 ≤ 3.0 or> 3.0	Arc - With power follow through
> 0.1 < 1.0	< 1.0	< 1.0	≥ 0.5 ≤ 3.0	Arc-No power follow through
≥ 1.0 < 3.0 or>3.0	< 1.0	< 1.0	< 0.5	Slight overheating - to 150°C
≥ 1.0 ≤ 3.0 or > 3.0	> 1.0	< 1.0	< 0.5	Overheating 150-200°C
> 0.1 < 1.0	≥ 1.0	< 1.0	< 0.5	Overheating 200-300°C
> 0.1 < 1.0	< 1.0	≥ 1.0 ≤ 3.0	< 0.5	General conductor over heating
≥ 1.0 ≤ 3.0	< 1.0	≥ 1.0 < 3.0	< 0.5	Circulating currents in windings
≥ 1.0 ≤ 3.0	< 1.0	≥ 3.0	< 0.5	Circulating currents core and tank; overload joints

Note: Several simultaneously occuring faults can cause ambiguity in analysis.

Table 11.4   Ratio Method of Diagnosis

11.6.4   Method for the Determination of The Gases Dissolved in Insulating Oils

The oil is sampled by means of a 250 ml capacity glass syringe. Details of sampling are found in clause 3.5, of IEC draft 10A (Secretariat) 25 "Guide for sampling oil from oil filled electrical equipment." (11.21) Details of other devices and methods of sampling of gases and oil are also given.

The gas is extracted from the oil by allowing the oil to flow into an evacuated vessel and then by repeated degassing by means of a Toepler pump most of the gases are extracted (Fig. 11.5). More details of the extraction apparatus and technique can be found in Clause 5.1.2 of IEC 10A (Secretariat) 25. (11.21) or (11.22)

Having extracted the gas from the oil, the gas is injected into a gas chromatograph equipped with two parallel columns packed with Porapak N and molecular sieve respectively. Each column is fitted with thermal conductivity and flame ionisation detectors. Further details can be found in IEC draft 10A (Secretariat) 25. (11.21) or ref. (11.22).

Transformers can contain considerable amounts of "fault" gases, depending on the number of years they have been in service, due to the natural ageing of insulation without necessarily the presence of severe electrical faults.

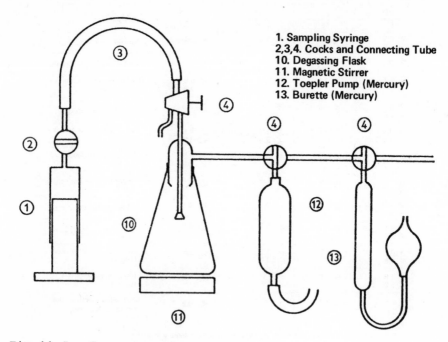

1. Sampling Syringe
2,3,4. Cocks and Connecting Tube
10. Degassing Flask
11. Magnetic Stirrer
12. Toepler Pump (Mercury)
13. Burette (Mercury)

Fig.11.5   Gas extraction apparatus

Table 11.5 gives an indication of the maximum concentration of these gases in oil at which a transformer may still be considered normal. These must be taken only as guidelines, depending on the age of transformer.

TABLE 11.5   Maximum concentration of dissolved gases in oil for transformers after many years service.

Gas	ppm(v/v)
Hydrogen	75
Methane	100
Ethylene	30
Ethane	30
Acetylene	15
Carbon Monoxide	500

Similar tables are given by Kelly (11.18) for the USA and ERA Ref.IE/418 (Ref.IE5/8) for Rhine Westfalia Electricity, Transformatoren Union AG, Germany and Laborelec, Belgium. (11.19)

The Transformatoren Union AG table gives values for transformers in service for less than 4 years, between

4 - 10 years and more than 10 years. The acceptable
concentrations of fault gases increase with the number of
years' service.

It is important therefore to monitor the state of the
insulation in a transformer on a routine basis by dissolved
gas analysis before a definite assessment can be arrived
at. Any increase in the concentration of fault gases over
a certain period can be approximately translated into energy
dissipated, provided the volume of oil is also known. It
has been reported that approximately 55 ml of gas is produced per kW Sec. at STP. (11.20) (11.17); from this the
intensity of the fault can then be estimated from the
calculation of the total dissolved gas found in the
transformer.

11.7   AGEING OF INSULATING LIQUIDS

11.7.1   Oxidation

Insulating fluids, especially those based on hydrocarbons are attacked by atmospheric oxygen in service to
produce oxidation products such as organic acids, water
and sludge. The rate of oxidation normally increases with
temperature(i.e. rate doubles/7°C rise). Tests have been
devised for the evaluation of the oxidation stability of
insulating oils; these are usually carried our in the
presence of O_2 at elevated temperatures in order to obtain
meaningful results in a reasonable length of time. The
oxidation stability can be assessed either by measuring
the time taken by a static film of the oil to absorb enough
quantity of O_2 at atmospheric pressure to produce an acid
value of 1 mg KOH/g or by determining the total amount of
oxidation products produced by bubbling oxygen or air
through the oil at various rates, in a given time and
temperature. These test can be carried out in the presence
or absence of a catalyst (usually copper).

In the first case (The Static Oxidation Stability Test)
the chemist has devised a method whereby 10g of oil is
heated in atmosphere of Oxygen at atmospheric pressure and
150°C until the oil absorbs 30 ml of pure oxygen. The time
taken to absorb this quantity of Oxygen is termed as the
useful life period (T300).

In the second case CERL (11.23) introduced a method
where oxygen or air is bubbled first through 25g of oil
held in an oxidation test tube for 164 hours and 100°C and
then through a similar tube containing distilled water at
room temperature to measure volatile acids (IEC 74), or
for a length of time to produce a total volatile acidity
of 0.28 mg KOH/g at 120°C (IEC474). In both cases, at the
end of the oxidation period, the oil sample is cooled and
diluted with n-heptane to precipitate solid oxidation
products. The oil solution is then filtered through a
sintered disc filter, the residue washed free of oil, dried
and weighed to determine the amount of insoluble sludge
produced. The acidity of the filtrate including the wash-
ings is determined to obtain the non-volatile acids

produced.

The addition of antioxidants and metal passivators have the effect of prolonging the period to produce the same amount of oxidation products (11.23). This is often achieved by the introduction of an induction period during which no perceptible change take place. The inhibition of oils should not however be used to enable poor quality oils to meet a specification, but its purpose should be to prolong the useful life of the oil and improve its electrical characteristics by reducing oxidation products.

Some inhibitors can get used up in short periods of time to produce sludge without the expected accompanying amounts of acidity (amine based inhibitors). Similarly unsuitable base oils can also sludge badly without the accompanying acidity in the presence of inhibitors.(11.23)

Under-refined oils when inhibited exhibit a pronounced induction period during which little oxidation occurs followed by a rapid rate of oxidation; such oils have very low aromatic and sulphur contents and would therefore require monitoring at frequent intervals. Normal and over-refined oils can oxidise, under low-oxygen conditions even when appreciable amounts of inhibitor are still present.

Oxidation tests for inhibited oils should be carried out for longer than the standard 168 hours test period used for uninhibited oils; a minimum test period of 504 hrs would be required to produce equivalent results with uninhibted oils, for test periods of 164 hour periods. Comparisons of inhibited and uninhibited oils can be made by IEC 474 oxidation method using an air flow of 0.15L/hr (low oxygen condition) rather than IL/hr (excess oxygen conditions for periods of 164 hours for uninhibited oils and multiples of 164 hours upto 504 hours). (11.23)

REFERENCES

11.1   BS148 (1972) "Specification for insulating oils for transformers and switchgear." Publication No. Gr. 7.

11.2   BS5730 (1979) "Maintenance of insulating oil". Publication No. Gr. 5.

11.3   IEC Standard "The determination of water in insulating oils". Publication 733.

11.4   Franklin, A.M. (1974)"Simpler measurement of relative humidity in electrical insulating oil". Electrical Review. 12 April.

11.5   Waddington, F.B. (1979) "Instruments to measure the salient properties of fluid dielectrics". IEEE Publication CH1510. EIC Meeting, Boston, October.

11.6   Redfearn, M.W. and Wilson, A.C.M. (1977) "The determination of aromatic compounds as a guide to the carcinogen activity of mineral oils. CEGB - RD/L/RI677.

11.7   BS4689 (1980) "Method for determination of flash and fire points of petroleum products: Cleveland open cup method." (ISO2592).

11.8   BS2000: Part 35: (1982) Flash point (open) and fire point of petroleum products by the Pensky-Martens apparatus.

11.9   Pipel, N. and Reynolds, E.H. (1960). "Modern dielectric materials." Heywood & Company Ltd., London, England.

11.10   Petersen, B.,(1968) Brown Boveri Review 55, No. 4/5 P222-227.

11.11   ERA 10A(1979) (Secretarial) 59, March

11.12   Cothen, H., (1972) "The Buchholz relay." 26: 131-136.

11.13   How, V.H., Massey, L. and Wilson, A.C.M.(1956) "Identity of gases collected in Buchholz protectors." M-V Gazette. 27: 139-148.

11.14   Domenburg, E. and Gerber, O.E. (1967) "Analysis of dissolved and free gases for monitoring performance of oil-filled transformers." Brown Boveri Review. 54: 104-111.

11.15   Vora, J.P. and Aicher, L.C., (1965) "Transformer fault-gas analysis andinterpretation." IEEE Winter Power Meeting, New York, N.Y., January 31 - February 5, 1965.

11.16   Waddington, F.B. and Allan, D., (1969). "Transformer fault detection by dissolved gas analysis." Electrical Review. 23May 751-754

11.17   Waddington, F.B. (1974) "Fault prevention in power systems by trace gas analysis: GEC I. Science Technology. 41:89-93

11.18   Kelly, J.J. (1980)"Transformer fault diagnosis by dissolved-gas analysis." IEEE Transactions on industry application, Vol. 1A-16, No.6

11.19   Interpretation of dissolved gas analysis: (1978) CIGRE papers by Dornenburg/Schober and E. Serena" ERA Reference IE/418 (Ref. 1E5/8).

11.20   Nichols Hazelwood, R., Frey, R.M.,Brocker, J.B., (1955) "Transformer oil. Properties of napthenic and aromatic fractions." Journal of the Electro-

Chemical Society 102 No.4

11.21   IEC draft 10A (Secretariat) 25: (1977)
"Guide for sampling of oil filled electrical
equipment and analysis of gases and of oils
for dissolve gases."

11.22   IEC Standard publication 567 (1977) "Guide
for sampling of gases and of oil-filled
equipment and for the analysis of free and
dissolved gases."

11.23   Wilson, A.C.M. (1980) "Oxidation of mineral
insulating oils "Insulating liquids: their
uses, manufacture and properties" 47-65.
Peter Peregrinus Ltd., Stevenage, U.K. and
New York.

Chapter 12

Thermal evaluation and
life testing of solid insulation

G. C. Stevens

12.1 INTRODUCTION

Electrical insulation is expected to withstand
electrical, mechanical, thermal and chemical stresses
without failure for at least 25 years. In the electrical
power industry this insulation is mostly organic and
susceptible to a deterioration in physical properties when
stressed. This is particularly the case when these
materials are thermally stressed. The successful application
of organic based insulation requires therefore a precise
knowledge of the thermal behaviour and thermal endurance of
the material and an assessment of the degree of degradation
it may suffer during the life of the equipment in which it
will be used.

To meet engineering needs, standards have evolved to
evaluate the long term thermal endurance of insulating
materials. These are currently presented in a collection of
International Electrotechnical Commission (IEC) publications
(12.1-4).The standards employ methods based upon accelerated
thermal ageing and considerable debate has occurred on
improving the reliability of these methods and in seeking
viable rapid-test alternatives.

This paper will introduce the concepts of thermal
ageing and the methods currently being recommended to assess
thermal endurance. Such conventional methods may be
usefully supported by the application of a number of physical
chemistry techniques. These complementary techniques will
be described and their use in evaluating thermal behaviour
and in providing a rapid prediction of ageing will be
discussed.

12.2 PHYSICAL AND CHEMICAL AGEING PROCESSES

Electrical insulating materials may consist of a
single 'simple' linear or network polymer or be produced
from composites involving organic or organic-inorganic
mixtures. Most materials are complex in their chemical
composition and their thermal behaviour. Several physical
processes may affect properties. Obvious examples are

softening or melting of materials above their glass
transition temperature or crystalline component melting
point. During ageing changes in chemical structure could
alter these transitions. Less obvious physical effects
include secondary crystallisation in partially crystalline
materials causing increased stiffness and structural
relaxation in an amorphous component causing densification
and embrittlement.

Potential chemical processes are more numerous and
interconnected. The most common deterioration routes
involve oxygen. Direct oxidation with or without the
formation of acidic groups will increase the conductivity
and dielectric loss. Oxidative enhanced chain scission or
crosslinking will dramatically alter mechanical properties;
scission leading to softening and crosslinking to hardening,
both with attendant changes in strength. Simple, thermally
induced (or pyrolytic) depolymerisation, chain scission and
crosslinking may also occur with the same consequences.
Water may attack a material through hydrolysis. Acid group
formation and increased ion-dissociation will influence
electrical properties and chain scission processes (e.g.
ester linkage reactions) may affect mechanical properties.

Additives are present in almost all insulation systems
and include plasticizers, extenders, fillers, antioxidants,
accelerators, drying and crosslinking agents, flame
retardants, and others. These will influence thermal
behaviour and physical properties during ageing either by
loss or by direct involvement in chemical reactions with
the host material. The latter is a particular problem with
as received materials entering service incompletely reacted.

12.3 THE ARRHENIUS RELATION AND THERMAL ENDURANCE

Thermal endurance testing makes direct use of the
empirical relationship commonly attributed to Arrhenius
which relates the rate of a thermally activated chemical
reaction (k) to inverse absolute temperature through the
equation

$$k = A \exp(-E/kt) \qquad \qquad \dots (12.1)$$

where A is a pre-exponential or rate factor and E is an
activation energy. The theoretical derivation is more
properly credited to H. Eyring who in 1935 (12.5) used
activated complex theory to obtain a statistical/
thermodynamic expression from which equation (12.1) is
obtained. It is clear from the derivation that equation
(12.1) only applies in the simplest cases (dilute gas or
solution reactions). In solids A is strictly temperature,
volume and entropy dependent.

Historically, V.M. Montsinger and T.W. Dakin provided
the seed leading to todays endurance test methods. Brancato
(12.6) has usefully reviewed this development. Montsinger
(12.7), in his 1930 paper on the thermal deterioration of the

tensile strength of transformer paper suggested an
empirical equation relating life τ to temperature in degrees
Celsius.

$$\tau = A \exp [-BT] \qquad \qquad \ldots (12.2)$$

A and B are constants characteristic of the material. This
equation provided a platform for the often quoted 10 degree
rule i.e. that the thermal life of insulation is halved
for each increase of $10^{\circ}C$ or doubled for each $10^{\circ}C$ decrease
in temperature.

    Montsingers improvement in predicting ageing was later
superseded by Dakins 1948 paper describing a chemical rate
treatment of thermal ageing (12.8). This forms the basis of
current thermal endurance test methods. Dakin assumed that
changes in a physical property P, should be related to the
concentration of an important chemical constituent of the
insulation C; more strictly

$$P = f(C) \qquad \qquad \ldots (12.3)$$

    Reaction rate theory (12.9) shows that the instantaneous
rate of change of C is proportional to some power of C such
that

$$\frac{dC}{dt} = -k \, C^n \qquad \qquad \ldots (12.4)$$

k is a rate constant and in this nth-order description of
rate processes n usually has a value between 0 and 3.
First-order reactions (n = 1) are readily integrated to give

$$C(t) = C_o \exp [- kt] \qquad \qquad \ldots (12.5)$$

Dakin argued that although deterioration processes rarely
followed first-order kinetics, it was generally possible to
find a f(C) which behaves linearly in time

$$f_o(C) = -kt \qquad \qquad \ldots (12.6)$$

For equation (12.5), $f_o(C) = \log_e(C)$. Similarly then,
given the assumption of equation (12.3) one can obtain for
a physical property a similar relationship to equation
(12.6)

$$f_o'(P) = -kt \qquad \qquad \ldots (12.7)$$

Combining equations (12.1) and (12.7) gives

$$\frac{f_0'(P)}{t} = k = A \exp(-E/T) \qquad \ldots (12.8)$$

If a constant value of $f_0'(P) = P_0$ is chosen to represent the life of the material at a point in time $t = \tau$ then

$$\frac{P_0}{\tau} = A \exp(-E/T)$$

and $\log_e \tau = \dfrac{E}{T} + \log_e P_0/A$ $\ldots (12.9)$

Therefore the logarithm of the time to reach a particular value of a property is proportional to the reciprocal absolute temperature. So a linear plot of the two should provide a reliable method of extrapolating higher temperature 'accelerated' measurements to lower temperatures close to the maximum service temperature proposed for the material.

12.4 CONVENTIONAL THERMAL ENDURANCE TESTING

IEC publication 216 provides guidance for the evaluation of the thermal endurance of insulating materials according to the principles leading to equation (12.9). The current parts of the existing standard (12.1-12.4) are detailed and brevity allows discussion of the key parts only.

Ageing behaviour is sensitive to the geometry and environment of the material under investigation. These should be chosen according to the engineering requirements and service conditions appropriate in practice. Similarly, the properties for study should reflect the engineering performance requirements. In many cases, more than one electrical or mechanical property is important. However, statistical considerations require that large numbers of specimens are aged and if destructive or disruptive tests are required, time and economic constraints may prohibit extensive endurance tests of all important properties. Preliminary work is therefore required to identify the most important factors when planning endurance tests which may extend over many months or years.

12.4.1 Data Generation and Graphical Construction

After the conditions of exposure and the properties to be monitored are prescribed the following steps are taken;

(1)   for each property of interest (e.g. tensile strength, flexibility, electric strength or loss tangent) an end-point is chosen corresponding to a limiting value of that property such that the material will still perform satisfactorily.

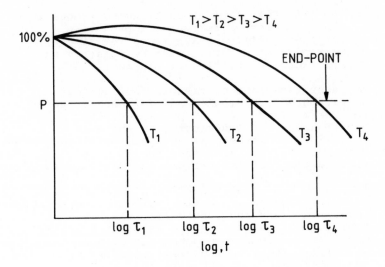

Fig. 12.1 Hypothetical property change during ageing at decreasing temperatures T_1 to T_4 and the choice of end-point

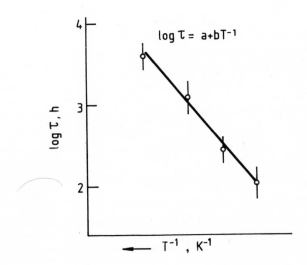

Fig. 12.2 Hypothetical thermal endurance graph generated from the end-point choice of Fig. 12.1

As shown in Fig. 12.1 at each ageing temperature
T_i the end-point criterion will produce a
corresponding 'life' value τ_i on the time axis.
Each property-time curve must be constructed
from a random population of samples.

(2)   at least three and preferably four ageing
        temperatures are studied and for each property
        the thermal endurance graph representing
        equation 12.9 is constructed as shown in
        Fig. 12.2. This data is analysed by linear
        regression ($y = a + bx$) and the a, b and
        correlation coefficients are calculated.

(3)   if a number of statistical requirements are
        met (3) (such as linearity, variance, correlation
        coefficient and confidence limits), three or
        four thermal indices may be calculated which
        describe the endurance behaviour of the
        material. These indices will clearly be
        end-point dependent and more than one set
        could be appropriate to any single property.

12.4.2   Thermal Endurance Indices

Four indices have been proposed which describe the
endurance results and allow classification and comparison
of materials. These indices may be understood by reference
to Fig. 12.3.
   (i) The Temperature Index (TI);
   is the temperature in $^{\circ}C$ derived from the thermal
endurance relationship at a given time, normally 20,000
hours. From the regression equation

$$TI = b/(4.301 - a) - 273, \,^{\circ}C \qquad \ldots (12.10)$$

   (ii) The Relative Temperature Index (RTI);
   provides a comparative estimation of the performance
of an unknown material with a well understood or reference
material. RTI is determined by plotting the endurance
data for the reference and unknown materials on the same
axes and comparing them at the time corresponding to the
recognised service or classification temperature of the
reference material. The temperature of the unknown material
at this time is the RTI.
(iii) The Halving Interval in $^{\circ}C$ (HIC);
   is a temperature interval in $^{\circ}C$ which corresponds to
halving of the time to achieve the end-point at the
temperature of the TI or the RTI. This is the modern
equivalent of Montsingers $10^{\circ}C$ rule. Strictly the halving
interval in inverse degrees absolute HIK is more correct
but is not favoured by some parties. From Fig. 12.3 it can
be shown that

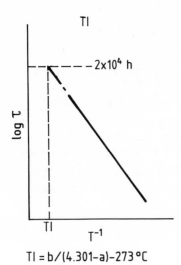

TI

$$TI = b/(4.301-a)-273\,°C$$

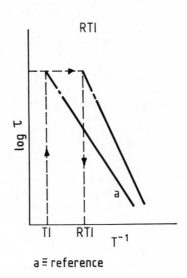

RTI

$a \equiv$ reference

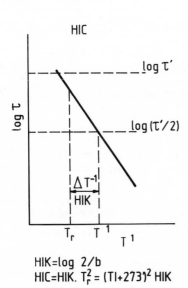

HIC

$$HIK = \log 2/b$$
$$HIC = HIK.\ T_r^2 = (TI+273)^2\ HIK$$

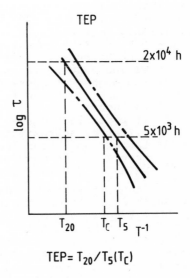

TEP

$$TEP = T_{20}/T_5(T_C)$$

Fig. 12.3   Derivation of the thermal indices TI,
RTI, HIC and TEP

$$\text{HIK} \quad = \quad \Delta T^{-1} (K) \quad = \quad \log 2/b \qquad \qquad \ldots \ (12.11)$$

and HIC is obtained from

$$\text{HIC} \quad = \quad T_r (K)^2 . \ \text{HIK} \quad = \quad (TI + 273)^2 \text{HIK} \ \ldots \ (12.12)$$

where the TI is chosen as the reference temperature for
convenience here.
(iv) Thermal Endurance Profile (TEP)
    Although the current part 4 of IEC 216 deals with TEP
(12.4) it may now be deleted in favour of TI and HIC solely.
TEP is a set of three temperatures in °C derived from the
endurance graph at the 20,000 and 5000 hour points with the
third obtained from the lower unilateral 95% confidence
limit on the temperature at 5000 hours.

$$\text{TEP} \quad \equiv \quad T_{20}/T_5 \ (T_c) \qquad \qquad \ldots \ (12.13)$$

    TEP is a very effective way of summarising thermal
endurance if the endurance graph is linear. As discussed
below, this is a requirement sometimes difficult to satisfy
making TEP generally unattractive.

12.4.3 Additional Time and Temperature Requirements

    To restrict extrapolation to reasonable limits
further requirements are recommended which apply to
preferably four exposure temperatures.
  (a) the lowest temperature should produce a median
      time to end-point of more than 5000 hours for
      TI and 2000 hours for RTI.
  (b) the highest temperature should produce a
      median time to end-point of not less than
      100 hours.
  (c) if possible the ageing temperatures should
      differ by approximately 20K.
  (d) the extrapolation necessary to obtain a TI
      should not be more than 25K and for RTI a
      value of 35K is acceptable.

    These requirements when combined with the statistical
acceptance conditions for the data provide safeguards for
erroneous application of the method. Nevertheless
intrinsic problems remain.

12.4.4 Limitations and Possible Remedies

    Many factors can influence the practical implementation
of endurance test methods. Some key contributory factors
include:

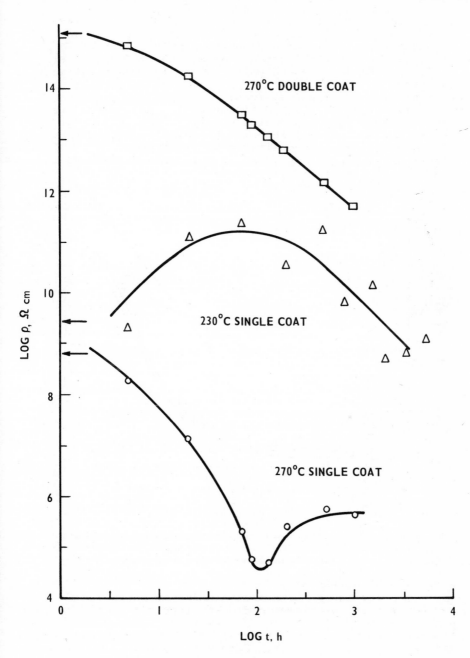

Fig. 12.4 Electrical conductivity behaviour of
a coating insulation material

Property Behaviour and End-Point Choice:
    linearity of the endurance graph will only occur if
the property of interest is simply related to a dominant
first-order degradation reaction in the material.   In
practice very few materials behave simply.   An example of
variable conductivity response on ageing is shown in Fig.
12.4.   In such a case end-point choice is a meaningless
exercise.
Multiple Degradation Reactions:
    more than one degradation reaction may contribute to
the behaviour of a physical property.   Competing or different
rates of reaction will invariably produce non-linearity in
the endurance graph even if the physical property is simply
related to the reactions.   An example of this for the
material of Fig. 12.4 is shown in Fig. 12.5.

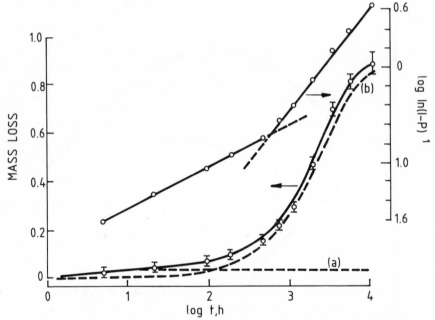

Fig. 12.5   Weight loss of the coating insulation
            of Fig. 12.4 at 230°C.   Symbols are
            experimental points, dashed lines are
            individual decomposition reactions and
            the full line is their sum.

Samples and Environment:
    testing large numbers of samples presents
reproducibility problems leading to inherent differences in
endurance behaviour.   Also, as reactions proceed the
physical and gaseous environment of the material may change
influencing its behaviour.   Such effects may be minimized
by changing conditions, if the problem is recognised.
However, the test may then not represent the service
situation.

These limitations lead to both non-linearity and high dispersion or scatter of endurance data. Scatter can be dealt with by the statistical treatment of outliers and testing additional specimens. Non-linearity can be accommodated by polynomial regression analysis. However, in order to provide reliable polynomial extrapolation it is necessary to generate lower temperature data. IEC suggests choosing an ageing temperature 10°C lower than the 5000 hour temperature to extend the time to end-point to around 10^4 hours.

These measures are required to retrieve unsatisfactory but not nonsense data. Clearly, a substantial investment in time and resources is required and the inherent limitations disuade many from thorough testing.

12.4.5 Thermal Classification

Assigning a thermal label or class to a material is attractive to design engineers who are shopping for suitable materials and equipment purchasers seeking a simple specification. Thermal endurance testing offers classification through the temperature index. However, for a classification scheme to remain useful as a simple aid it cannot fully categorise the vast number of engineering situations and environments encountered in practice. Indeed, the philosophy behind endurance testing recognises the need to appraise individual circumstances; this negates the concept of universal classification. That is not to say that classification for well prescribed cases is impractical, it may be useful (e.g. magnet wire insulation (12.6). Suitable choice of insulation for a particular task must remain an engineering judgement based on experience, record of performance or thorough investigation.

12.5 THERMAL ANALYSIS METHODS

The limitations of conventional thermal endurance testing arise primarily from its inability to both assess the underlying mechanisms responsible for ageing and to evaluate thermal behaviour.

A number of physical chemistry techniques can be used to provide a detailed understanding of thermal behaviour (12.10) and complement existing endurance testing. These techniques are summarised in Table 12.1. They may be used in three related areas:

Thermal Behaviour Characterisation

$\updownarrow$

Reaction Identification

$\updownarrow$

Kinetic Analysis and Prediction

A detailed account of each technique is not possible here. Currently the most useful techniques fall broadly into mass and energy analysis methods.

TABLE 1. Mass and Energy Analysis Techniques for Studying Thermal Behaviour

TECHNIQUE	SENSITIVITY	INFORMATION AVAILABLE
Mass Analysis		
mass spectroscopy (MS)	$10^{-12}-10^{-13}$ % min^{-1}	identification of decomposition products and mechanism, kinetic
gas chromatography (GC)	$10^{-9}-10^{-12}$ % min^{-1}	parameters at a molecular level, normally isothermal
combination GC/MS	10^{-12} % min^{-1}	analysis.
liquid chromatography	$10^{-6}-10^{-9}$ g	
infra-red spectroscopy	$10^{-5}-10^{-7}$ g	mass loss and gain analysis, bulk kinetic
thermogravimetric analysis (TGA)	10^{-3} % min^{-1} (10^{-6} g)	parameter evaluation, reaction separation, isothermal and nonisothermal analysis.
Energy Analysis		
differential thermal analysis (DTA)	10^{-2} mW s^{-1}	endothermic/exothermic reaction energy, specific heats, kinetic
differential scanning calorimetry (DSC)	10^{-3} mW s^{-1}	parameter evaluation, reaction separation, isothermal and
isothermal differential calorimetry (IDC)	10^{-3} mW g^{-1}	nonisothermal analysis
STA (combination TG/DTA/DSC)		
Miscellaneous		
thermomechanical analysis	50 ppm	dimensional analysis, viscoelastic and modulus parameter analysis under isothermal and nonisothermal conditions, some kinetic analysis possible.

The most sensitive techniques are mass spectroscopy (MS) and gas chromatography (GC) which analyse the gas phase products of decomposition. These products can be identified and their concentrations evaluated. Infra-red spectroscopy provides chemical group identification and concentration information in the solid and gas phase by vibrating groups absorbing infra-red light. Thermogravimetric analysis (TGA) simply measures bulk mass loss or gain. Of these techniques TGA has received most attention in providing ageing information and will be dealt with here. The other mass methods have received about equal attention largely in support of ageing studies. GC has recently been applied more systematically to ageing studies and will also be discussed.

All of the energy analysis methods rely on the heat absorbed or evolved when chemical reactions or phase transitions occur. Differential thermal analysis (DTA) detects this by monitoring the temperature difference between the sample and a stable reference. The calorimetry methods of Table 12.1 evaluate the amount of energy required to maintain a specified temperature (e.g. through a correction current). Of these techniques, very powerful combinations of TG with DTA (and potentially differential scanning calorimetry (DSC)) are possible and have been given the name simultaneous thermal analysis (STA); STA will be discussed here.

Paloniemi has made extensive use of isothermal calorimetry in thermal endurance testing. However, it is applied in an appraisal of endurance based on the concept of equalising thermal ageing rates; this is achieved by manipulating the environment. Space does not permit discussion of this topic and the reader is referred to a three part review of the technique (12.11).

These techniques can complement conventional testing in several ways.
(1)   Through thermal behaviour and reaction identification to:
  (a) provide a technique based comparison of thermal stability and an alternative classification.
  (b) guide endurance test planning.
  (c) provide rapid material screening.
(2)   Through kinetic analysis to:
  (d) set the conditions of endurance tests rapidly.
  (e) construct life curves which may be correlated with or referenced by conventional life curves.

Translating this complementary information to the engineering situation is possible but can present difficulties. This will be illustrated by the use of examples.

12.6   QUALITATIVE THERMAL BEHAVIOUR

Although substantial information can be obtained from individual techniques, combinations have the advantage of working on the same sample undergoing change in the same thermal and gaseous environment. Commercial simultaneous thermal analysers currently employ a TGA/DTA combination but the qualitative interpretation of a TGA/DSC combination is the same. The quantitative parameters obtained under linear temperature scanning (nonisothermal mode) are illustrated in Fig. 12.6. These provide a basis for both thermal and chemical characterization and comparative thermal stability (12.12),(12.13) in any gaseous environment. The thermograms provide a 'fingerprint' of the material and its behaviour.

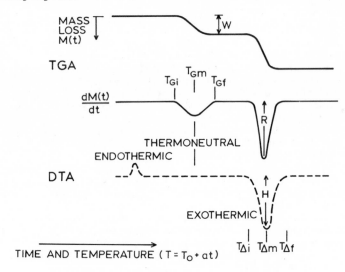

Fig. 12.6   Nonisothermal analysis thermogram parameters such as onset (i), maximum (m) and finish (f) temperatures in TGA (T_G) and DTA (T_Δ) are used qualitatively. Parameters such as mass loss (W) and rate (R) and heat of reaction (H) may be used quantitatively

Crosslinked polyethylene (XLPE) provides a useful example particularly when compared with its low density polyethylene (LDPE) pre-cursor. Its STA behaviour in nitrogen and air is shown in Fig. 12.7. At $105^{\circ}C$, DTA indicates a melting transition in XLPE which is reproduced in air. In LDPE melting occurs up to $10^{\circ}C$ higher. DSC analysis indicates a broad melting behaviour beginning at $60^{\circ}C$ in both polymers and reveals distinct crystallinity differences. In air around $200^{\circ}C$ weight gain due to oxidation, which occurs exothermically, is apparent. This reaction occurs at higher temperatures in LDPE. This is followed by two decomposition reactions in air for XLPE but only the first one occurs in LDPE. These occur through the formation of hydroperoxides which reflect the chemical structure differences in the two polymers. In contrast, in nitrogen higher temperature endothermic reactions are observed. In LDPE these correspond to crosslinking and chain scission occurring around 200 and 300°C with scission dominating above 290°C. The latter becomes increasingly effective at higher temperatures such that low molecular weight gases are rapidly ejected above 350°C resulting in weight loss. In LDPE in nitrogen this decomposition behaves as a single reaction in STA. In XLPE this weight loss occurs up to 60°C lower and two stages of decomposition are seen. These principal reactions in air are oxidatively enhanced and occur 60 to 100°C lower in temperature and proceed very exothermically producing sample self heating. In Fig. 12.7 this has lead to spontaneous ignition. At higher temperatures the remaining carbon residues remain stable in nitrogen but are exothermically 'burnt off' as CO and CO_2 in air.

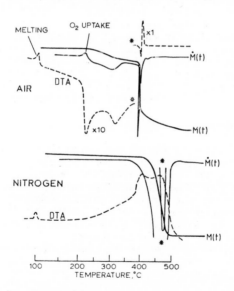

Fig. 12.7   Simultaneous thermal analysis of crosslinked polyethylene in air and nitrogen

If an endurance test was planned for this material some relevant points could be raised.
(i)   crosslinked polyethylene undergoes partial melting at relatively low temperatures when it will suffer a drop in its mechanical moduli. Will properties measured at room temperature after ageing truly represent service behaviour?

(ii) oxidation and subsequent degradation in air are complex multi-reaction events highly dependent on initial polymer structure and oxygen availability. Are the endurance test specimens representative of the service situation?

(iii) competitive crosslinking and chain scission reactions occur and precede significant weight loss. Can an ageing temperature range be chosen to include only one reaction or should both be included?

Such information is helpful and improves our understanding of the material. However, clear answers to some of the questions above may not be forthcoming and compromise may be required. The important point is that we are in an informed position to make a judgement.

12.7 QUANTITATIVE KINETIC ANALYSIS

In principle the kinetics of change of any parameter may be followed usefully provided a mathematical description is available. Evolved gas analysis and mass loss analysis both sample the consequences of a decomposition reaction at a molecular level and should be amenable to a kinetic analysis based on the Arrehenius relationship (equation 12.1). If engineering properties of interest relate to the same chemical processes leading to mass loss, life information could be more rapidly assessed by a non-conventional route.

12.7.1 Gas Chromatography

The sensitivity of GC is sufficiently high to detect reactions occurring at the rates expected at service temperatures or below where ageing occurs over tens of years. The technique may complement endurance tests by (12.13).

(i) determining the rate of gas evolution at various temperatures and times and plotting this rate against inverse temperature; this essentially provides the slope of the endurance graph.

(ii) determining one point of the conventional endurance graph at a single high temperature (short test) for the properties of interest.

This is a minimum requirement which relies on the respective apparent activation energies of the gas and property being identical.

Epoxy resin and polyethyleneterephthalate (PETP) provide examples where CO and CO_2 feature as the principal evolved gases resulting from oxidative decomposition reactions. The resulting rate-graphs are shown in Fig. 12.8. The epoxy resin exhibits the linear behaviour ideally required for an endurance test whereas PETP exhibits obvious nonlinearity. Other work indicates that both materials undergo oxidative decomposition but PETP also undergoes hydrolysis which begins to dominate above $180^{\circ}C$. Transforming the PETP data onto a thermal endurance graph (by taking the time for 1 mole of $CO + CO_2$ to appear from

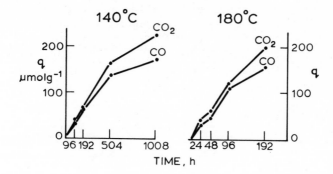

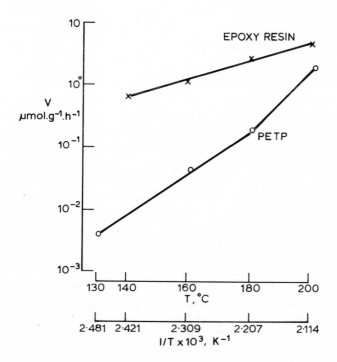

Fig. 12.8   Rate of CO + CO₂ evolution for
epoxy resin and polyethylene-
terephthalate (PETP)

1 g) gives the comparison of Fig. 12.9 where a conventional endpoint of 2% elongation to break was chosen. The agreement between slopes and nonlinearity is satisfactory.

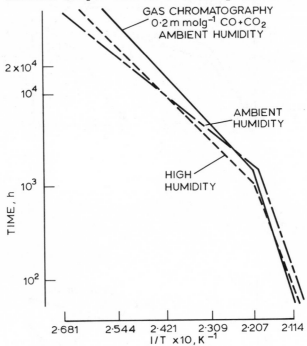

Fig. 12.9   Comparison of the conventional
            endurance and gas chromatography
            thermal graph for PETP

12.7.2   Thermogravimetric Kinetics

        TGA and DTA have been used to kinetically predict the thermal endurance of insulating materials; a good example is Toop's work on wire enamels (12.14). In TGA, kinetic parameters may be obtained from isothermal experiments and the philosophy applied to the GC method above can be used. A second, more rapid alternative is to obtain the linear rate heated nonisothermal thermogram and kinetically analyse this. In general a wide variety of kinetic models may be used to describe these thermograms (12.15-12.17). However, a simple n-th order kinetic model with Arrhenius rate constants has been found to describe most results effectively.
        It is possible using iterative methods to obtain the reactant mass and rate constants of each mass loss reaction contributing to a multi-reaction decomposition. From these the time to achieve a particular mass loss from either a single controlling reaction or from a group of reactions can be calculated. An example of this is a coating insulation whose conventional endurance behaviour is shown in Fig. 12.5.

STA indicated two principal first-order decomposition
reactions whose reactant masses were dependent on preparation
conditions. Conventional testing showed that properties
such as coating electrical resistivity, bond strength and
film thickness paralleled mass loss. A mass loss criterion
of life is then acceptable. Choosing 63% of reaction (a)
and 33% of the total mass available for decomposition
provides two mass loss criteria representing minor and
major deterioration in properties. The endurance curves for
these conditions obtained from nonisothermal TGA are shown
in Fig. 12.10. A number of conventional endurance and
isothermal TGA results are also included. The agreement
is acceptable when allowance is made for the differences in
material and test conditions.

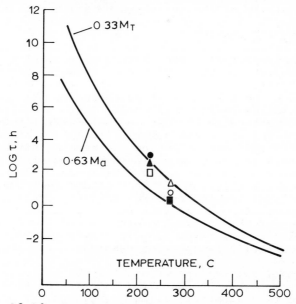

Fig. 12.10   Coating insulation thermal graphs.
Comparison of nonisothermally determined
thermal endurance (full lines) with
that determined isothermally and
by conventional testing.
Symbols;
($\square$) 230°C and ($\blacksquare$) 270°C isothermal
in N_2 for 0.63 M_a, ($\bigcirc$) 0.09 M_T =
0.63 M_a and ($\triangle$) 0.33 M_T endurance
test in N_2 at 270°C, ($\blacktriangle$) 0.09 M_T
endurance test in N_2 at 230°C,
($\bullet$) 230°C manufacturers endurance
test in static H_2 (0.09 M_T for
double coat)

TGA methods cannot match the sensitivity of GC and extrapolation of the kinetics to times and temperatures outside those experienced experimentally is required. However it is an integral technique which in combination with energy analysis provides several interpretational routes to complement endurance studies. Recently, Flynn and Dickens (12.18) have introduced the technique of factor-jump thermogravimetry to assess activation energies from differential mass loss observations. Interestingly for a number of solids Dickens has shown that the activation energy is dependent on the extent of the reaction (12.19). This result illustrates the care required in interpreting any thermally activated behaviour, particularly along routes that assume the constancy of kinetic parameters or the applicability of the Arrhenius relationship.

12.8   CONCLUDING REMARKS

Conventional endurance testing although limited, allows a complete thermal endurance appraisal of the engineering situation of interest. Likewise, the engineering properties of most relevance to the service life of the insulation can be monitored. Thermal analysis and other complementary techniques may not fulfill these requirements and it is not always simple to translate complementary information to the engineering situation. Currently this can only be achieved by comparison with conventional test results or by empiricism.

However, the investment in time and resources required for conventional testing requires that the options presented by the complementary techniques be fully exploited. In materials development this approach is most attractive where comparative or relative performance can be established quickly to identify prospective candidates for more detailed study.

Conventional testing is unlikely to be replaced totally by newer methods but the detailed information they offer should lead to a substantial improvement in our understanding of the thermal behaviour and endurance of solids.

REFERENCES

12.1   IEC Publication 216, Guide for the determination of thermal endurance properties of electrical insulating materials: Part 1. General procedures for the determination of thermal endurance properties, temperature indices and thermal endurance profiles., 1974 (Also BS 6591: Part 1, 1979).

12.2   Part 2. List of materials and available tests, 1974, (Also BS 5691: Part 2, (1979).

12.3   Part 3. Statistical methods for the determination of thermal endurance profile, (1980).

12.4   Part 4.   Instructions for calculating the thermal
       endurance profile, (1981).

12.5   H. Eyring, J. Chem. Phys 3 107 (1935); also
       S. Glasstone, K.J. Laidler and H. Eyring 'The
       Theory of Rate Processes', McGraw-Hill,
       New York (1941).

12.6   E.L. Brancato, 1978, 'Insulation Ageing, A
       Historical and Critical Review', IEEE Trans.
       Electr. Insul. EI-13 No. 4, 308.

12.7   V.M. Montsinger, 1930, 'Loading Transformers
       by Temperature', AIEE Trans 49, 776.

12.8   T.W. Dakin, 1948, 'Electrical Insulation
       Deterioration Treated as a Chemical Rate
       Phenomenon', AIEE Trans 67, 113.

12.9   K.J. Laidler, 1969, 'Theories of Chemical
       Reaction Rates', McGraw-Hill, New York.

12.10 'Thermal Methods of Analysis', 1974, Vol. 19
       of Chemical Analysis W.W. Wendlandt, John
       Wiley, New York; and 'Thermal Characterisation
       of Polymeric Materials', 1981, Editor:
       E.A. Turi Academic Press, New York.

12.11 P. Paloniemi, 1981, 'Theory of Equalisation of
       Thermal Ageing Processes of Electrical
       Insulating Materials in Thermal Endurance Tests.
       Parts I, II and III, IEEE Trans. Elect. Insul.
       EI-16(1), 1.

12.12 G.C. Stevens, 1979, 'Simultaneous Non-isothermal
       Analysis in the Study of Dielectric Materials';
       DMMA Conf. Univ. Aston.

12.13 G.C. Stevens, B. Fallou and A.G. Day, 1982,
       'Complementary Techniques in the Thermal
       Endurance Testing of Electrical Insulating
       Materials', CIGRE paper 15-05.

12.14 D.J. Toop, 1971, three papers in IEEE Trans.
       Elect. Insul. EI-6 (1) 2-14, 25-36.

12.15 J.H. Flynn and L.A. Wall, 1966, 'General
       Treatment of the Thermogravimetry of Polymers',
       J. Res. Nat. Bur. Stand. 70A (6), 487.

12.16 C.D. Doyle, 1966, in 'Techniques and Methods of
       Polymer Evaluation', Vol. I, Thermal Analysis'
       Editors, P.E. Slade and L.T. Jenkins, Chapt. 4,
       Marcel Dekker, London.

12.17   J. Sestak, V. Satava, W.H. Wendlandt, 1973,
        <u>Thermochimica Acta</u>. <u>7</u>(5), 447.

12.18   J.H. Flynn and B. Dickens, 1976, <u>Thermochimica
        Acta</u> 15, 1.

12.19   B. Dickens, 1979, three papers in <u>Thermochimica
        Acta</u> <u>29</u> 41-113.

<u>ACKNOWLEDGEMENT</u>

Part of this work was undertaken at the Central Electricity Research Laboratories and is published by permission of the CEGB.

Chapter 13

Evaluation of micaceous composites

C. S. Clemson

13.1   INTRODUCTION

Electrical grade mica is still one of the best electrical insulating materials, having a high volume resistivity (10^{11} - 10^{15} ohm.m) a high electric strength (>10^5 Vmm^{-1}) and excellent resistance to erosion by prolonged discharge. Mica splittings bonded with natural gums have been used commercially in electrical insulation from the start of the electrical industry, nevertheless micaceous insulating materials are still being developed, notably in the form of micapaper products bonded with a variety of modern synthetic resins.

Following a brief review of the general properties of micas and micaceous products examples of the methods of evaluating flexible tapes for the stator winding insulation of large AC generators will be discussed, also rigid laminates for commutator separator plates and heater plates.

13.2   MICAS AND MICACEOUS PRODUCTS

Micas (13.1, 13.2, 13.4) are naturally occurring minerals, widespread in the earth's crust, they are all related in structure and properties. Micas suitable for electrical insulation - MUSCOVITE and PHLOGOPITE are expensive and some what rare and the majority of such micas are mined in India.

Micas are formed from complex silicate structures consisting of layers of oxygen atoms grouped around silicon, aluminium and other metal atoms and are weakly held together by layers of widely spaced potassium atoms.

The best electrical micas are clear to transparent, varying in degrees of colouration (muscovite generally ruby or green, phlogopite amber) poorer grades have inclusions or spots which may represent electrical weaknesses and are dark coloured. Muscovite is especially notable for its perfect cleavage and exceedingly thin elastic sheets which can be easily split from the crystal. Good muscovite has a low loss tangent, typically in the range 10^{-3} - 10^{-4} over a wide band of audio and radio frequencies, with a permittivity in the range 5 - 7, density 2.6 - 3.3 g.cm^{-3} and hardness MOH 2.5 - 3.2.

Micaceous products (13.1, 13.2) are manufactured in the

form of rigid sheets, tubes and mouldings also as flexible
sheets and tapes.

Built-up micaflake materials in the form of sheets and
tapes have been widely used in high-voltage insulation; the
material is manufactured by laying-up splittings, either by
hand or machine, with overlap in one plane and bonding them
with shellac, bitumen etc.

A moulded mica product is manufactured from powdered mica
and borosilicate glass heated to a semi-molten condition
(700°C). The material is specially good at high frequency
and has high voltage withstand also high resistance to moisture and high temperatures.

Although single aggregates of muscovite crystals up to
85 tonnes in weight have been found (13.4) nevertheless the
supply of large sheets is limited and attempts have been
made to alleviate this situation by the following developments :

13.2.1 Synthetic Mica

Synthetic Mica (13.1) has been prepared in which the
hydroxyl groups are replaced by fluorine. This material
has never competed in price with the natural product.

13.2.2 Micapaper

Micapaper (13.1, 13.6, 13.8, 13.9) is made wholly from
treated mica scrap. The paper is made on plant like that
used for the manufacture of ordinary paper. The mica plate-
lets are about 1μm thick with a wide range of diameters,
maximum 1mm diameter; they are held together by VanDer Waals
cohesive forces which gives the paper limited strength.
The paper is available in thicknesses from 0.05mm to 0.23mm
in continuous lengths and its quality can be controlled, its
uniformity is an essential characteristic for its application in electrical insulating products.

13.3 APPLICATION OF FLEXIBLE GLASS-BACKED MICACEOUS
TAPES IN GENERATOR BAR INSULATION

An important use of mica (whether built-up or micapaper)
has been for the insulation of high voltage machines, notably the main insulation of large generator stator windings
(13.1, 13.3). In many cases a Class B (130°C) temperature
rating is necessary, but even when this is not so (where
conductors are water cooled for example) mica is used because of its endurance under discharge conditions.

13.3.1 Bitumen - bonded Micaflake Tape Insulation

Glass backed, bitumen-bonded micaflake tape insulation
was used for many years over the whole range of output, up
to 200 MW AC generators and the largest DC machines. The
glass backing is primarily required to give the tape sufficient tensile strength to enable it to be handled during
manufacture and in its application. The generator insulation

is made by winding on to the conductor stack the flexible
micaflake tape (Fig.13.1) to a thickness of about 4mm for
an 11kV machine. (13.1).

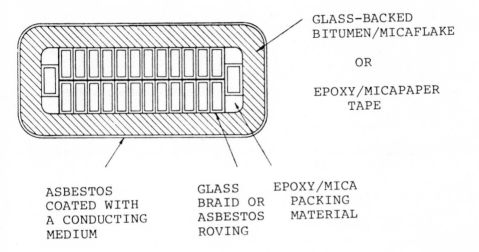

GLASS-BACKED
BITUMEN/MICAFLAKE

OR

EPOXY/MICAPAPER
TAPE

ASBESTOS
COATED WITH
A CONDUCTING
MEDIUM

GLASS
BRAID OR
ASBESTOS
ROVING

EPOXY/MICA
PACKING
MATERIAL

FIG.13.1   Cross-section of slot portion of
typical alternator coil (13.11)

With increasing size of generators this situation changed in
the 1950's (13.1, 13.6) when it was found that on long machines the thermoplastic bitumen/micaflake system was no
longer practical because difficulties occurred due to long-
itudinal movement of the tape arising from differences in
thermal expansion between copper and insulation which caused
cracks to appear in the wall of the insulation at the end of
the stator slot ('tape-separation'). In addition continuing
demand for greater output from a given frame size required
finding space for more copper in the slot and operating at
higher temperature rises and high mechanical forces (due to
bar vibration). As a result the thermoplastic bitumen insulation is no longer used in large machines and it has been
replaced by synthetic resin systems.

13.3.2   Thermosetting Resin-bonded Micapaper Insulation

Micapaper by itself has little mechanical strength until
it is treated with a resin and the properties of the final
product are obviously influenced by the choice of resin.
Most of the products formerly made from built-up mica can be
reproduced from micapaper impregnated with various resins

and bonded under pressure and heat. In the case of glass backed tape the choice of glass cloth (13.7, 13.8), usually about 0.04mm thick and the nature of the finish affects the properties of the insulation. Typically the cloth is woven from E glass fibres treated with a silane finish.

For a resin to be suitable for long term operation (13.9) it requires :

a)   High thermal stability with low electrical loss at the service temperature and at power frequency.

b)   Excellent adhesion to mica, backing materials and copper.

c)   Good resistance to electrical discharge (corona).

d)   High resistance to moisture, chemicals and other contaminants.

e)   High mechanical strength over a range of service temperatures.

f)   Thermosetting properties, giving dimensional stability.

g)   The capability of being used in insulating tapes for hand or machine application.

h)   The ability to operate at Class F temperatures (155°C).

i)   A short cure time at 150°C to 160°C.

After preliminary tests the only system to find favour in the UK is the Epoxy-Novolak (EN) range of resins. The fullest descriptions of such systems are given by PARRISS (13.7) and SMITH (13.5). Epoxy-Novolak resin was selected (13.5) because when it is used in conjunction with a boron-trifluoride complex catalyst it can be processed readily to a stable, flexible, B stage (a soft semi-cured resin condition) which is required for ease of handling in the taping process. On further heating under pressure it will compact and cure to a thermally stable hard resin, without the evolution of volatile by-products - an essential property if void-free insulation is to be produced with a low pressure moulding (1-2 Nm^{-2}) technique.

There are two fundamentally different methods of manufacturing stator windings :

13.3.2.1   The Resin-rich Process

In this process the tape is liberally impregnated beforehand with EN resin in a volatile solvent (typically acetone or MEK) and dried to a slightly soft B stage (solvent concentration about 0.7%) with enough flexibility to be bent without cracking. The retained solvent has a plasticizing

effect. Catalyst concentration determines the gel time,
which must be long enough to allow sufficient resin flow to
occur to give good consolidation of the insulation. The
tape can be hand or machine wound onto the conductor.
After taping the bar is hot pressed to consolidate and partially cure it (typically 45 mins. at 160°C) the excess resin
being extruded together with trapped air.

13.3.2.2 The Vacuum-Pressure Impregnation Process (VPI)

The VPI process involves the application of dry, or low
resin content, tape onto the stator coil. The insulated
coil is vacuum impregnated with a liquid resin and subsequently cured under pressure.

High quality stator windings of all types have been
developed using the 'resin-rich' EN bonded micapaper material. The uniform nature of the basic tapes, combined with
their excellent insulating properties means that modern
windings are inherently more reliable than those made in
the past using shellac and bitumen 'thermoplastic' bonding
agents. The 'resin-rich' system has been used on the largest sizes of high voltage machinery, in particular turbine
generators with voltage ratings up to 30kV.

13.3.3. Test Data on Epoxy Novolak bonded Micapaper Insulation

The principle properties of EN micapaper/woven glass
laminate are recorded in Table 13.1 (13.12). These laminates were cured for 1 hour at 160°C followed by 16 hours
at 140°C post cure.

ELECTRICAL			
	Thickness		Test Method
Dielectric Strength at 20°C - MV/m (Normal to Laminate)	1mm 3mm 5mm	70 50 45	I.E.C.371-2 App. A.10
	Temp.		Test Method
Dissipation Factor 50 Hz 1 MV/m	20°C 130°C 155°C	0.006 0.029 0.070	I.E.C.371-2 App. A.11
	Temp.		
Dielectric Constant 50 HZ 1 MV/m	20°C 130°C 155°C	5.0 5.2 5.6	
			Test Method

Comparative tracking index -
                Glass side:
                Mica side:
		110
		190
			BS5901:1980 (I.E.C.112: 1979)

MECHANICAL/PHYSICAL			
Flexural Strength Mpa	Temp. 20°C 155°C	370 280	Test Method I.E.C.371-2 App. A.9
Flexural Modulus Gpa	Temp. 20°C 155°C	60 43	Test Method I.E.C.371-2 App. A.9
Water Absorption 2.5 thick	% mg	0.1 6.0	Test Method BS2782:1970 502F
Coefficient of Linear Expansion		$9x10^{-6}/°C$	
Thermal Classification		Class F 155°C	Test Method I.E.C.371-2 App. A.13

TABLE 13.1   Typical properties of EN resin/
micapaper/woven glass laminate (13.12)

The data illustrates the excellent electrical strength
of the laminate and its low loss characteristics. The low
loss tangent tip-up obtained in production coils for an
18kV machine are illustrated in FIG.13.2 (13.8)

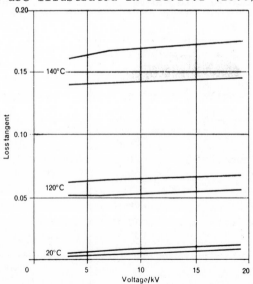

FIG.13.2   Relation of loss tangent to voltage at
various temperatures measured on stator
bars insulated with EN resin/micapaper
insulation on one machine (13.8)

Comparison data for tan δ as a function of voltage and of temperature for stator coils insulated (a) with shellac - bitumen micaflake (b) VPI epoxy micapaper and (c) resin-rich epoxy micapaper are illustrated in FIGS.13.3A & 13.3B (13.9).

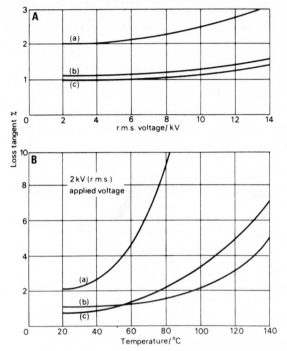

FIGS.13.3A,B   Graphs of loss tangent against voltage (A) and change of dielectric characteristics, with temperature (B) for stator coil, insulated with (a) shellac/bitumen micaflake (b) impregnated epoxy micapaper (c) resin-rich epoxy micapaper (13.9)

Since a high level of loss tangent of an insulating material may engender high temperatures in the material under service conditions and eventual dielectric failure, it is obvious that epoxy mica systems are superior to the bitumen micaflake system.

The dielectric strength of the new material is about 40% higher than that of bitumen material, so that the use of the former permit a 20% reduction in slot insulation thickness, which in turn promotes better heat transfer.

13.3.4   Test data on model stators wound with epoxy novolak bonded micapaper insulation

In addition to material tests on the basic epoxy novolak micapaper system a series of long-term tests have been reported on stator bars and on model stators (13.8, 13.9) as follows :

13.3.4.1 High Voltage Ageing Programme

About 700, 80cm long stator bars were insulated with eight different insulation systems to an 11kV specification (13.9) and aged in ovens at several temperatures in the range 90°C to 150°C. The samples were continuously stressed at 2kV mm⁻¹ for the first 20,000 hours and subsequently at 3kV mm⁻¹. The change in loss tangent of each sample between 4kV and 15kV (defined as the loss tangent tip-up) was calculated. FIG.13.4 compares the loss tangent tip-up curves for a resin-rich epoxide bonded micapaper system and shellac/bitumen bonded flake mica, aged at 130°C.

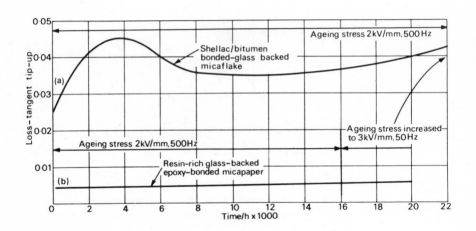

FIG.13.4   Graph showing variation of loss tangent
            tip-up with time for sample bars at 130°C
            under continuous electric stress of 2kV mm⁻¹
            & 3kV mm⁻¹ (a) Shellac/bitumen micaflake
            (b) resin-rich epoxy micapaper (13.9)

13.3.4.2 Thermal Cycling Test

A segment of a model stator core was wound with hollow copper 22kV coils, insulated with both resin-rich and VPI epoxy micapaper and flake mica insulation systems, to form a thermal cycling rig (13.9). The model stator was energized at phase voltage, the coils being heated electrically to 130°C and then cooled by pumping water through them. The rate of 3 cycles per hour achieved was much faster than that experienced in service and the total number of cycles completed by the rig was equivalent to 20 years of normal

service life. No tape separation, nor any significant
changes occurred in the insulation systems.

13.3.4.3 Bar Vibration Test

Slot vibration forces which could lead to the delaminat--
ion of the insulation become significant in machines generating 500 MW and above and the effect of such vibrations was
examined in another stator model, using both bitumen thermoplastic and epoxy thermosetting resin insulated bars (13.8).
The greater amplitude of vibration in the bitumen insulated
bars can be seen in FIG.13.5

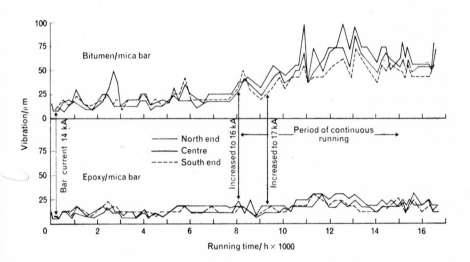

FIG.13.5 Comparative vibration levels of thermoset
and thermoplastic insulation systems
measured on the bar vibration rig (13.9)

13.3.4.4 Model Stator Tests

To simulate the forces experienced by an insulation
system under short circuit conditions a 22kV, 500 MW stator
replica was built (13.9) with full size end windings but a
short core. The current corresponding to an instantaneous
three-phase short circuit at the generator terminals was
injected and the end winding movements were recorded by
position transducers and high speed cinematography. These
photographs have been used to assure manufacturers of the
basic suitability of the insulation material on a complete
stator winding system.

13.3.5 Quality Control Tests

During the manufacture of windings stringent quality
control techniques are used to monitor all critical aspects
of the process (13.9). Basic insulating materials are purchased to rigid specifications. All bars and coils are subjected to various electrical tests during manufacture and
before being wound into the stator core. The most important
tests for checking consistency of manufacture are the loss
tangent/voltage characteristic measured with a Schering
bridge and the electrical discharge characteristics which
are computed with the aid of a Dielectric Loss Analyser.

13.3.6 Site Testing

Methods of assessing the condition in service of HV
rotating machine insulation are similar to those used in
quality control (13.10). Visual inspection reveals tightness of
wedges etc. and possible presence of electrical discharge
damage. Loss tangent/voltage tests, using a Schering bridge,
reveal voids in the insulation. Discharge energy measurements using a Dielectric Loss Analyser supplement the loss
tangent measurement in that they show the discharging of
voids in the insulation and thus a measure of the lack of
consolidation of the insulation. These electrical tests
have been designed to determine the overall condition of the
insulation. If a 'weak' area exists it will not be shown by
the above tests. A method of detecting weak areas employs
a 0.1 Hz Test set, which is an easily portable equipment
which simulates quite closely the 50 Hz stresses imposed on
the winding in service. Sound stator insulation, even in an
'aged' condition will as a general rule withstand an applied
voltage of 4 X rated line voltage. If there are any localised 'weak' areas the application of 1.5 X rated machine
voltage for one minute enables them to be found.

13.4 OTHER APPLICATIONS OF MICACEOUS LAMINATES

13.4.1 Commutator Separator Plates

These are insulating barriers between adjacent copper
segments of rotating machine commutators. The conventional
material is micaflake sheet bonded with about 5% shellac
resin. Micapaper bonded with an epoxy resin system has been
found to have considerable advantages over the conventional
material (13.9).

Commutators are built to very close dimensions. The
number of separators used in a given commutator must give a
precise 'height' under consolidated conditions of the commutator. For this reason the separators are 'stack gauged':
the height of a given number of separators, measured under
load of 7 MPa (1000 lbf in^{-2}) must fall within specified
tolerances.

Micapaper laminates, which are very uniform, can be pressed to close tolerances and by selection of such sheets very
fine limits can be obtained. The following tests have been

performed on epoxy bonded micapaper commutator plates (13.9):

13.4.1.1 Behaviour under Compression

Although a finalized commutator must be of a very stable construction and no slippage of copper can be tolerated, nevertheless during building, which embodies successive tightenings of the assembly, a very slight yield of the insulation under applied loads is beneficial. By careful formulation and processing of an epoxy micapaper system characteristics the same as those of the conventional shellac/micaflake material can be obtained.

A given stack of separators, from whose edges any burrs are carefully removed, is subjected to repeated compression at $160°C$ of loads of 7 MPa and 55 MPa and the stack height is measured after each application of load. The cycling and stack height measurements are repeated at least 5 times. Test data are shown in FIG.13.6, for a stack whose initial height was 17.8mm.

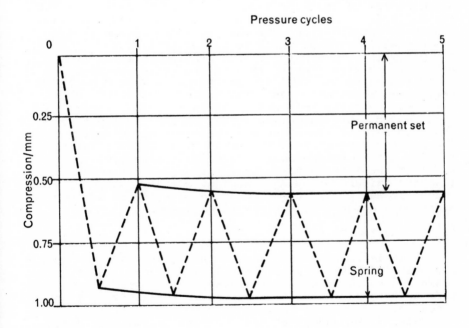

FIG.13.6 Load/compression cycling diagram for epoxy-bonded micapaper commutator laminate at $160°C$.

The remaining 'springiness' or resilience is the property which has been adjusted to give a close match for the conventional shellac/micaflake material and is of the order of 2.5% of the initial stack height. Commutator manufacturers, however, design on the overall change in stack height

after the initial bedding load has been applied; hence a
more meaningful figure is obtained by taking the combined
permanent set and residual springiness and calculating the
percentage change based on the initial stack height at 7MPa.
This figure is of the order of 5.5%, to match micaflake/
shellac material.

13.4.1.2 Flexural Strength

When the ends of the rotor coil are attached to a commutator it is common practice to bend the riser copper some
what to allow easier entry of the coil ends. The flexural
strength of flake material is of the order 100 MPa while
that of the epoxy micapaper material is much higher at
345 MPa. However, if the distortion is too great the micapaper risers could be snapped off and care must be taken to
work within acceptable limits.

13.4.1.3 Electrical Tests

Comparative Tracking Index tests were carried out on
conventional flake material and a range of micapaper sheets
and in almost all cases CTI was beyond the scope of the
commercial instrument which recorded up to 700V.
One method of arc-resistance testing is to arrange for
an arc to be struck every 10 secs. and to force this arc
downwards on to the test surface by means of a powerful magnet, a 6mm stack of separators, one edge of which was sanded
flat, was used for this test. Again a range of products
were tested. All materials withstood 1000 arcs without
track formation. The test was therefore not discriminating,
but did not suggest that the epoxy micapaper was inferior
in edge arcing to the shellac micaflake material.
From these tests there is no strong reason to expect
tracking failure of epoxy micapaper commutators in service.
Indeed it has proven to be the case in practice and epoxy/
micapaper commutators are now widely used.

13.4.2 Heater Plates

Rigid laminates are used for insulating high temperature
heaters in domestic and industrial appliances (13.1). The
resin system is generally silicone, but other high temperature systems have been considered, including fully inorganic
types.
Some of the more pertinent properties examined during
the development of a satisfactory material were as follows
(13.9) :

13.4.2.1 Effect of Temperature

Inorganic bonds will be largely destroyed when the heater
plate is taken to 600°C and some products will be gaseous,
which will tend to form blisters. This cannot be tolerated
to any great extent and it is considered that a bonded system should pass the following test :

A 100 x 50mm sample of laminate carried on a light non-shielding frame is placed in a muffle furnace controlled to 600 - 625°C. The sample shall not be closer than 75mm to any wall of the muffle. After one hour at the above temperature, the sample shall be removed and examined after cooling to room temperature. There shall be no blistering nor delamination of the sample.

Only those materials which passed this test were considered further.

13.4.2.2. Hardness of the Laminate

In many instances the sheet of heater plate would be wound with wire, there is a strong tendency for the wire to cut into the edge of the sheet. A simple test was devised to give a comparative numerical rating of the 'edge hardness' of the heater plate.

A loop of 18 s.w.g. nickel-chrome wire, attached to a stirrup, is placed over a 50 x 16mm sample supported to within 6mm of a long edge. Weights are carefully added to the stirrup so that the wire cuts into the edge of the sheet under test. The end point is taken as the load required to cause the wire to penetrate the sheet edge to a depth equal to the wire diameter and can be judged quite well by a light sweep of the finger tip along the edge of the sample. Ten readings are taken and the average figure quoted.

With good quality micapaper the individual results are remarkably consistent. Test data are presented in Table 13.2.

13.4.2.3 Mechanical Strength

Since there is generally a mechanical requirement, a mechanical strength test is important in assessing the various types of material available. It was felt that flexural strength was most instructive. Test data on various thicknesses of laminates are recorded in Table 13.2.

13.5 CONCLUSIONS

Micaceous insulation is still predominant in applications requiring high discharge resistance and/or high temperature resistance.

Micapaper has been available for many years, but because the tiny flakelets seemed more susceptible to damage than the well-tried mica flakes, many of the potential users had initial doubts which long-term trials have now eliminated.

The material has steadily been introduced into applications ranging from small domestic appliances to the largest turbo-generators, and as confidence has been established, its use has started to increase rapidly to the exclusion of micaflake products.

The author wishes to thank Dr. G.F. Smith, the Micanite & Insulators Co. Ltd., for helpful discussions during the preparation of this review of Micaceous materials.

PHYSICAL PROPERTIES	THICKNESS/ mm	RESULT	TEST METHOD
Flexural strength at 20°C/MPa	0.18 0.38 0.51	95 110 88	BS2782– 304B
Blister test at 600°C		No blistering after 1 hour	See Text
Electrical strength/MV m^{-1}	0.18 0.38 0.51	44.8 40.8 41.2	BS2782– 201C
Comparative Tracking Index		> 700	BS3781
Edge hardness/g mm^{-1}	0.18 0.38 0.51	3700 5590 6535	See Text
Water absorption/%, mg	0.18 0.38 0.51	6.8, 72 5.6, 116 5.3, 137	BS2782– 502G
Density/g ml^{-1}		1.8	BS2782– 509A
Resin content %		10	BS626-A

TABLE 13.2   Properties of silicone resin bonded micapaper heater plate material.

REFERENCES

13.1   Sillars R.W. 1973 "Electrical Insulating Materials
        and their Application" IEE Monograph 14. Peter
        Peregrinus Ltd. Stevenage, UK.

13.2   Molloy E. (Ed) 1947 "Electrical Materials" George
        Newnes, London.

13.3   Walker J.H. 1979 "Large A.C. Machines, Design,
        Manufacture and Operation" Bharat Heavy Electricals
        Ltd., Thomson Press, New Delhi, India.

13.4   Sinkankas J. 1966 "Mineralogy : A First Course"
        D Van Nostrand, Princeton, U.S.A.

13.5   Smith G.F. 1980 "Modern Mica Insulation Systems"
        International Electrical Insulation Seminar,
        Paper 1 B3 Bombay.

13.6   Farmer E. 1962 "Some Modern Trends in Electrical
        Insulation Materials" Research 15. 54 - 62.

13.7   Parriss W.H. 1971 "Material requirements for Epoxy-
        Novolak bonded High Voltage Machine Insulation" GEC
        J.Sci and Tech. 38, 4, 157.

13.8   Ashworth E. and Murdoch C.O. 1974 "Properties of
        Micapaper as an Electrical Insulation Material"
        GEC J. Sci & Tech. 41, 4, 107 - 116.

13.9   Greenwood P. and McNaughton H.S. 1972 "Insulation-
        Keynote to Machine Reliability" Energy International
        9, 8.

13.10   GEC Machines Ltd. "Insulation Testing of Machines"
        Publication 3609 - 1 Rugby UK.

13.11   Farmer E. 1970 "The Insulation of Large Electrical
        Machines" BEAMA Elec. Insulation Conf. Paper 1, 131.

13.12   The Micanite & Insulators Co. Ltd. 1983 "Mimica
        EN2" Publication MP3, Manchester UK.

Chapter 14

Techniques for electrical non-destructive evaluation of materials

W. P. Baker

14.1 INTRODUCTION

When an electric field is applied to an insulating
material, the response of the dielectric can be a reversible
one in which case the field can be applied and removed many
times, or the response can be irreversible, for example the
material may puncture or decompose. The following notes
refer solely to reversible responses, and to their measurement.

What purpose can be served in making such measurements?
The answers to this question can be divided into three
groups. In the first the measurements can be used as a
control on the consistency of quality of materials that are
being continuously used in the production of standard
manufactured items so that there can be reasonable confidence that the quality of the finished items will all meet
a standard specification. Secondly, short term measurements are made to try to appraise the long term value of
new insulating materials. This is much the most difficult
task of dielectric measurement. The third group is
concerned with measurements designed to discover what has
gone wrong to cause a failure of a piece of insulation in
service.

The measurements are directed to quantifying one or
more of the four basic properties of an insulating material
namely permittivity, conductance, loss tangent and
dispersion.

The permittivity cannot normally be measured directly
but the relative permittivity (ε_r) can be deduced from
capacitance measurements since ε_r = Capacitance/C_o where
C_o is the capacitance between the electrodes with the
insulating material replaced by vacuum.

It is generally found that ε_r is not a constant for
most materials but that it varies with time of stressing,
or under a.c. conditions with frequency. This fractional
variation with frequency is defined as dispersion, from
the optical analogy. Under a.c. conditions the feature of
reversibility is found not to be realised perfectly and the
shortfall or imperfection can be described in terms of loss
tangent ($\tan\delta$).

Again tanδ is found not to be constant but to vary with frequency and more particularly with temperature. An obvious contribution to this imperfection is the conductivity (G) of the material, apparently a manifestation of a continuous conducting path between the electrodes.

The measurement of dielectric properties is the measurement of these four properties.

It is found that the measurements fall naturally into two main groups. For simple materials ε_r (deduced from capacitance) and tanδ may be measured as functions of frequency or of temperature and the magnitude of the applied voltage need be little more than that required to make the measurements easy to carry out. For insulation systems, on the other hand, as may be found in complete equipment, transformers, motors generators and so on, for example, the main interest may well be the performance as a function of voltage and the measurements would be made at constant frequency and perhaps over a small range of temperature. The choice of frequency and temperature falls naturally to the working conditions of the piece of equipment, generally 50Hz and the temperature range 20°C to 100°C.

Nearly all measurements on capital plant fall into this latter category; it will be considered first.

The most extensively used measure of an insulating material is loss tangent because it has the particular merit, as a measure of the quality of a specimen of insulation, that it is dimensionless; direct comparisons can consequently be made on similar materials having widely different geometries.

14.2 THE SCHERING BRIDGE

The first well-established means of measuring capacitance and loss tangent at high voltages was the Schering bridge shown in its simplest form in Fig. 14:1.

The conditions of balance may be readily derived if the lossy test-specimen is represented by a capacitance X in series with a loss-resistance r. The other arms of the bridge are: S, a standard high-voltage pure capacitor; a variable resistor R_3; and a (usually) fixed resistor R_4 connected in parallel with a variable capacitor C_4. Applying the balance condition that the product of opposite-arm impedances must be equal, then:

$$R + 1/j\omega X = (R_3/R_4 j\omega S)(1 + j\omega C_4 R_4) = R_3(C_4/S) + R_3/j\omega S R_4 \tag{14.1}$$

Hence, by separating real and imaginary terms:

$$R = R_3(C_4/S), \quad X = S(R_4/R_3), \quad \text{and} \quad \tan\delta = \omega X r = C_4 R_4 \omega \tag{14.2}$$

Apart from the condition for r, it can be seen that R_3 occurs in the expression for X, only and C_4 in the expression for tanδ only so that these two may be made

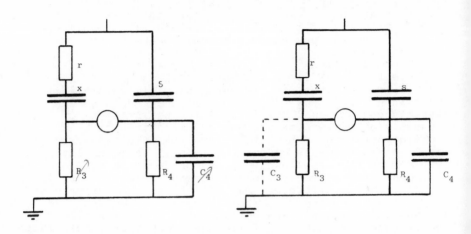

Fig.1   Simple Schering
        bridge

Fig.2   Scheringbridge
        with C_3

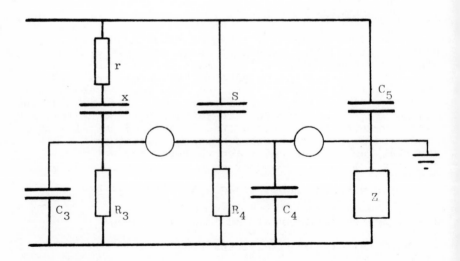

Fig.3   Six arm bridge

variable to achieve a bridge network in which two elements are independent of each other. This is one merit of the network; another is that the two variable elements are in the low-voltage arms and may be operated with safety even though the total applied voltage may be several hundred kilovolts.

The physical realisation of the bridge network must depart a little from the simple circuit of Fig. 14:1 because the bridge cannot be compact with high applied voltages, and low-voltage leads must be screened. The main difference is the additional capacitance included across the low voltage arms as shown in Fig. 14:2. For this new circuit the arms are $(r + 1/j\omega X)$, $1/j\omega S$, $R_3/(1 + j\omega C_3 R_3)$ and $R_4/(1 + j\omega C_4 R_4)$. For balance:

$$(1 + j\omega Xr) \frac{X}{S} \cdot \frac{R_3}{R_4} \cdot \frac{1 + j\omega C_4 R_4}{1 + j\omega C_3 R_3} \tag{14.3}$$

giving:
$$\frac{1 + j\omega Xr + j\omega C_3 R_3 - \omega^2 Xr C_3 R_3}{XR_3} = \frac{1 + j\omega C_4 R_4}{SR_4} \tag{14.4}$$

Writing $\tan\delta = \omega Xr$ for the test specimen, also $\tan\delta_3 = C_3 R_3$ and $\tan\delta_4 = C_4 R_4$ and then separating real and imaginary parts, we obtain

$$X = S(R_4/R_3) (1 - \tan\delta_3 \tan\delta) \tag{14.5}$$

and
$$(\tan\delta + \tan\delta_3)/XR_3 = \tan\delta_4/SR_4 \tag{14.6}$$

which, using the relation already found for X, leads to

$$\tan\delta = \frac{\tan\delta_4 - \tan\delta_3}{1 + \tan\delta_4 \cdot \tan\delta_3} \tag{14.7}$$

The term $\tan\delta_3$ may easily be of the same order as the loss tangent of the specimen being measured, and it may not be easy to evaluate because C_3 is made up not only of screened-lead capacitance but also of stray capacitance distributed over R_3. Stray capacitance across arms 1 and 2 is found to be much easier to evaluate or eliminate than that across arm 3. Many attempts have to be made to reduce the significance of unknown components of C_3 and a number of these methods will be described.

14.2.1 The Wagner Earth

So far it has been assumed that the low-potential terminal of the supply transformer is connected to earth. Let us now examine a six-arm bridge as shown in Fig. 14.3. When the bridge is balanced all the detector terminals are at earth potential. The capacitances between these terminals and earth, having no voltage across them, do not now affect the balance conditions of the bridge. In addition to the detector leads, the low-potential leads making connection to the supply transformer have now to be

screened; the capacitances between these leads and the
screens are in parallel with the impedance of the sixth
arm - again not contributing to the conditions of balance
of the four main arms. The high-voltage capacitor making
arm 5 does not have to be provided as a separate unit
because it already exists in the original four-arm bridge
to earth, that is, by the incorporation of the Wagner earth.
Of course, the low-potential terminal of the supply
transformer must be lightly insulated, but the tank itself
should be earthed.

It is not usual to provide two detectors; the one is
switched from one position to the other in turn as the
adjustment of the three low-voltage arms converges to the
full balance condition.

The detailed nature of the Wagner earth impedance Z
has not yet been discussed. We find that the arbitrary
high-voltage capacitance of arm 5 is not necessarily an
advantage. To provide a phase balance in the simple four-
arm bridge, a capacitance C_4 has to be added across R_4 if
the loss-tangent of the specimen is greater than that of
the standard - as it almost invariably is. If the loss-
tangent of the specimen were lower than that of the
standard then the phase-shift of R_4 would require to be
reversed, i.e. an inductor would have to be added.
Although this condition seldom arises, a closely similar
condition exists between the specimen arm and the stray
capacitance arm, for the latter includes the transformer
high-voltage bushing and some of the winding insulation.
This capacitance is, therefore, far from loss-free. It
often has a higher loss-tangent than the specimen and
inductance is required in the arm in series with it. A
fixed inductor, astatically wound, can be free from
spurious induced voltages but the variable inductance
required for the Wagner earth is very difficult to make
free from induced voltages and is often a source of
trouble in power-frequency bridges.

14.2.2 Screen Driving

It is not the capacitance C_3 itself that introduces
errors into the bridge network but the current flowing
into that capacitance. In the Wagner earth system the p.d.
across, and consequently the current in, C_3 is reduced to
zero. In one other system a third electrode is interposed
between the core and outer screen of the cables
contributing to C_3 and C_4 and if this third electrode or
inner screen is driven at the same potential as the core,
the current flowing from the core is reduced to a very low
value.

14.2.3 Double Ratio-Arm Bridge

Another way of eliminating errors introduced by stray
capacitance is to balance the bridge twice with different
resistance values on the ratio arms but with the same
stray capacitance across them. By suitable choice of

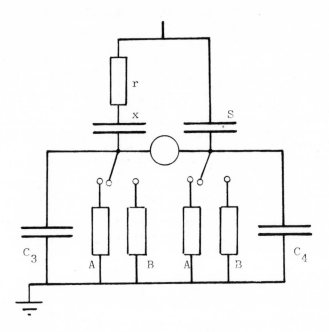

Fig. 14.4 Martin and Patterson bridge

resistance values, a compound expression for the loss-
tangent can be obtained eliminating C_3. An example of this
technique has been described by Martin and Patterson (14.1)
who used a variable standard capacitor in order to have
equal ratio arms ($R_3 = R_4$) for each of the balance
conditions. Referring to Fig. 14:4 for the first balance
$R_3 = R_4 = R_A$ say and $C_4 = C_A$ for the second balance $R_3 =
R_4 = R_B$ and $C_4 = C_B$.

$\tan\delta = \omega R_A(C_A - C_3) = \omega R_B (C_B - C_3)$ from which $C_3 = (R_A C_A -
R_B C_B)/(R_A - R_B)$

substituting for C_3 gives $\tan\delta = \omega R_A C_A - \omega R_A$

$$\frac{R_A C_A - R_B C_B}{R_A - R_B} \tag{14.8}$$

which simplifies to $\tan\delta = \dfrac{(C_B - C_A)R_A R_B}{(R_A - R_B)}$ (14.9)

With this arrangement the loss-tangent is obtained
from the simple expression and the time required to balance
the bridge is much shorter than that for a six-arm bridge.
A drawback is that the applied voltage is limited to about
1kV, the maximum for which variable standard capacitors
are available.

14.2.4 Other Errors of Schering Bridge Ratio-arms

In order that R_3 may be adjusted it usually takes the form of a six-decade variable resistor. The maximum value of R_3 is limited to $10,000\Omega$. It is unlikely that the total contact resistance of 6 switches in series can be as low as 0.01Ω and may well be several times this value. An accuracy of three significant figures can only be maintained, therefore, for values of R_3 above 10Ω, giving a range of capacitances that can be measured to this precision of about 1000 to 1. A shunt circuit for R_3 to enable higher capacitances to be measured more accurately will be described later. The need for high quality, and hence massive, switches endows them with large stray capacitance; further the ratio-arm assembly occupies a panel of perhaps half a square metre. The two ratio arms and the null indicator form a closed loop, possibly embracing enough stray magnetic field to introduce a spurious signal into the null indicator. This pick-up may be detected by observing the change in the balance condition when the ratio-arms are turned into several positions.

14.2.5 The Measurement of Large Capacitances

In order to maintain a high value of R_3 it may be shunted when large capacitances are to be measured. It is usual to add a resistor in series with R_3 as shown in Fig. 14.5(a) to avoid excessive current should R_3 be set inadvertently to a low value.

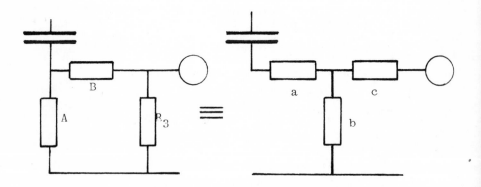

a   b

Fig. 14.5 R_3 Shunt

The balance condition is most easily found by converting the delta arrangement of R_3, shunt resistor A and series resistor B, Fig. 14:5(a), into the equivalent star shown in (b). The conversion gives "a" in series with the specimen, b as the new effective R_3 and c in series with the null indicator, where

$$a = AB/\Sigma R, \quad b = AR_3/\Sigma R, \quad \text{and} \quad c = BR_3/\Sigma R$$

and $\Sigma R = (A + B + R_3)$. The capacitance of the specimen is therefore $X = S/(R_4/b) = S(R_4/R_3) \left[1 + (B/A) + (R_3/A) \right]$

$$(14.12)$$

The resistance a in series with the specimen increases the overall loss-tangent by

$$\omega X a = \omega S R_4 (a/b) = S R_4 (B/R_3) \tag{14.13}$$

The effective resistance c in series with the null indicator does not influence the balance condition. It has negligible effect on the sensitivity if the null indicator is voltage sensitive, i.e. if it has a high input impedance. On the other hand if the null indicator is current-operated or has a matching transformer input, then the sensitivity may be reduced, at worst, by two or three times.

14.3 INDUCTIVELY-COUPLED RATIO ARMS

The improvements in magnetic core materials in the last few decades have enabled inductive potential-dividers and current-comparators to be made with high precision. It is natural that this development should be exploited by the designers of measuring equipment and there is now a variety of bridges available for the measurement of admittance or impedance over a wide range of frequencies, all based on transformer coupling of one sort or another.

Two particular bridge networks are of interest in the measurement of insulation properties. One, due to Lynch uses a transformer as a voltage source; the other, developed originally by Blumlein (14.2) and more recently by Glynne (14.3) uses a differential transformer as a current-comparator.

The latter, shown in Fig. 14:6 is more suited to measurements at high voltages and power frequencies, particularly if the number of turns of one ratio arm can be varied. At balance, indicated by zero voltage induced in the search coil there is no flux in the common core and hence no net m.m.f. produced therein by coils 1 and 2: the balance condition is therefore $I_x N_1 = I_3 N_2$, where N_1 and N_2 are the numbers of turns in series with X and S respectively. The title 'ampere-turn ratio' applied to this bridge follows from the balance equation.

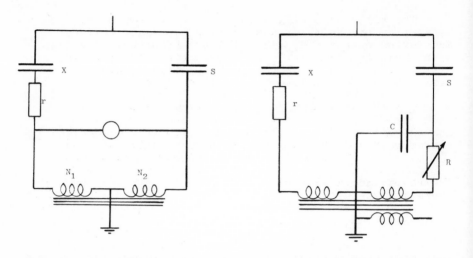

Fig. 14.6 Blumlein bridge   Fig. 14.7 Glynne bridge

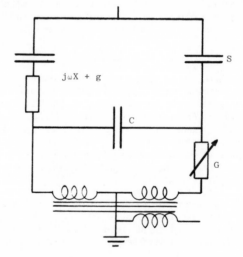

Fig. 14.8 Baker bridge

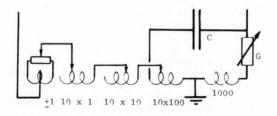

Fig. 14.9 Baker bridge with variable N_1

Zero net m.m.f. not only results in zero search-coil voltage but zero voltage across all windings, a characteristic which gives the ampere-turn ratio arms their particular merit. The capacitances across the windings, arising from screened low-voltage leads do not enter into the balance condition because there is no voltage developed across them. Long low-voltage cable runs can be used with these ratio-arms without a Wagner earth or other artifact. The sensitivity of the bridge can be increased to a value many times that of the Schering bridge which which has to be made insensitive to minimise the loss-tangent increments, $\omega R_3 C_3$ and $\omega R_4 C_4$, due to stray capacitance. It is often stated that the sensitivity of a transformer bridge is twice that of a resistance-capacitance network. This is true only if the latter has arms of roughly equal impedance but it is a prime requirement of the high-voltage bridge that the supply voltage is very unequally divided.

The simple bridge of Fig. 14:6 can balance pure capacitance only; it is necessary to add some means of providing a phase balance. That due to Glynne is shown in Figure 14:7. The current I_S in the standard capacitor S is divided by C and R into two components in quadrature, so that the current I_2 in the N_2 turns of coil 2 for unit applied voltage is

$$I_2 = I_S/(1 + j\omega CR) = 1/ \mid R/(1 + j\omega CR) + (1/j\omega S)(1 + j\omega CR) \mid$$

$$= j\omega S/ \mid 1 + j\omega(S + C)R \tag{14.14}$$

The balance equation is consequently:

$$\frac{N_1 j\omega X}{1 + rj\omega X} = \frac{N_2 j\omega W \ (G + j\omega C) \ G}{G + j\omega(S + C) \ (G + j\omega C)} \tag{14.15}$$

from which $X = \dfrac{N_2}{N_1} \times S$ and $N_1 \ (S + C) \ X = N_2 SGr\omega X$ (14.16)

but $\tan = Xr = \omega(S + C)/G$

if $1/G = R +$ winding resistance $= R + a$
$\tan\delta = \omega(S + C) \ (R + a)$ (14.17)

Hence the lowest value of specimen $\tan\delta$ that can be balanced is:

$$\tan\delta_{min} = \omega(S + C)a$$

Provision for capacitance and phase balance can be obtained independently when N_1 and R are made variable. The resistance R can be calibrated to read loss-tangent directly when the standard S is fixed.

The range of loss-tangent readings can be extended by switching capacitors into position C. If N_2 were made variable instead of N_1, there would appear to be the advantage that N_2 could be calibrated to read the capacitance of X directly; this is possible, but (as will be seen from the discussion of errors) it would be also necessary to maintain the d.c. resistance of N_2 constant as the number of turns was changed.

For the measurement of specimens of low loss tangent particularly less than the value given by $\omega(S + C)a$ there is a preferred modification to the Glynne bridge in which the phase retarding capacitor taking some of the current from the standard capacitor is connected across the ratio arms as shown in Fig. 14:8. The capacitor then serves the additional function of phase advancing capacitor adding to the current from the test specimen.

The balance condition is most easily derived by a delta star conversion in which:

$$\alpha = \frac{-\omega^2 SC}{g + j\omega \ (S + C)} \quad \text{and} \quad \beta = \frac{j\omega SG}{g + j\omega \ (S + C)} \qquad (14.18)$$

γ need not be evaluated since it takes zero current.

for balance: $$j\omega X + G - \frac{\omega^2 SC}{g + j\omega(S + C)} = \frac{j\omega Sg}{g + j\omega \ (S + C)}$$
$$(14:19)$$

from which $X = S - \dfrac{G}{g} (S + C)$ and $\tan\delta = \dfrac{\omega}{g} (2C + S)$ very

nearly substituting for $\tan\delta$ gives $X = S(1 - \frac{1}{2} \tan^2\delta)$
$$(14:20)$$

This value for X lies midway between the equivalent series value and the equivalent parallel value and is therefore more meaningful than either. Furthermore the $\tan\delta$ balance can be taken down to zero. When N_1 is made variable, as it must be to balance the capacitance with a fixed value of standard the bridge takes the form shown in Fig. 14:9.

14.3.1 Errors in the Inductively Coupled Ratio Arms

The simplest ratio-arms are wound as a three limb transformer. With a core of commercially available Mumetal about 5cm^2 in cross-section the fractional leakage reactance will be less than 1 part in 50,000.

In the equations of balance the d.c. resistance of the windings has been ignored whereas in fact, there will be a small voltage drop in the winding due to the resistance R_o. The magnitude and phase of the current in the bridge arms will be changed because of this voltage drop and the change in the phase angle will be the greater. The phase-angle error will be $\omega(XR_o - SR_o)$ and if either ωXR_o or SR_o is not to exceed 10^{-5} then R_o must not exceed 30Ω for X or S equal to 1000pF. This value can always be achieved, and the resulting modulus error for a phase-angle error of 10^{-5} is 10^{-10} which is quite negligible.

A much more serious source of error in the inductively

coupled ratio arms is induced voltages in the windings.   It
is easy to ensure electrostatic screening but more
difficult to provide adequate magnetic screens. The
reduction of a magnetic field by a mild-steel tube can be
computed from a formula given by Starling(5).

A mild steel tube 15cm long, 15cm in diameter and with
a wall 0.5cm thick would reduce the internal field to about
4 percent of the external field. (14.4).

14.4 NULL INDICATORS

A good null indicator should have adequate sensitivity
freedom from noise, selectivity and a sufficient bandwidth.
These requirements, difficult to achieve in the past, are
now a trivial design problem with modern micro electronics.
Detailed consideration is not now considered to be
necessary.

14.5 MEASUREMENTS AS A FUNCTION OF FREQUENCY

It was implied earlier that the capacitance of the
test specimen was of less interest than the loss tangent.
This arose because the geometry is often ill-defined and it
is not easy to calculate a value for permittivity that is
particularly meaningful. However, if the interest is in
the measurements obtained, not on insulation systems in
high voltage plant, but on simple specimens of dielectric
material, the value of permittivity becomes much more
important and it is then necessary to carry out the
measurement in a manner that minimises the errors in the
calculation of permittivity.

Such measurements made to characterise the material
need to be carried out over a range of frequency and
temperature. The choice of frequency variation is either
upwards through the AF RF and UHF ranges or downwards to
the ULF and ELF ranges. These different directions
required very different techniques and must be discussed
separately. Historically the frequency range has been
explored upwards following developments in Radio Technology
there is some logic, then, in describing this first.

14.5.1 Audio Frequency Measurements

Again, a choice is presented. Some measurements will
be required as a routine on a series of similar test
specimens and the main requirement is convenience in
measurement and a minimum of computation to evaluate
permittivity and loss tangent; this demands a direct
measurement of capacitance and loss tangent.

14.5.2 Direct Bridge Methods

Can the bridge networks described for use at 50Hz be
used at audio frequencies? Note the errors that increase
with frequency and the conclusion is that with a Schering
bridge a Wagner earth use of the bridge above 1kHz is
difficult. The ampere-turn type of inductively coupled

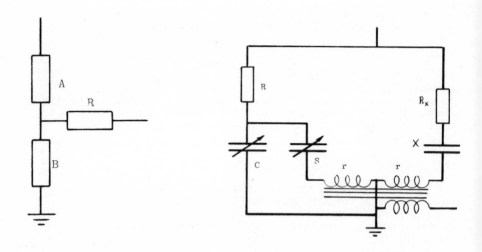

Fig. 14.10 Conductance arm   Fig. 14.11 AF bridge

bridge is perhaps a factor of ten better than the Schering
bridge in the low audio frequencies but it can be used up to
100kHz if specially made with a trifilar wound, fixed ratio
arms with null indicator winding and used in conjunction
with a variable standard capacitor. About three fixed ratio
arms are necessary to cover the range from 1kHz to 100kHz.
The tanδ balance described for use at 50Hz has to be
replaced with a variable standard conductance arm which can
usefully be made up of three resistors disposed as shown in
Fig. 14:10.

The lower arm of the potential divider B is made the
variable unit and must be of small physical size to minimise
the error introduced by its stray capacitance C_B. The magnitude of this error is $\omega C_B(A + B)$ and this must be smaller
than the lowest value of tanδ to be measured. The value of
conductance introduced in parallel with the standard
capacitor is:

$$G = B/(BR + RA + AB) \qquad (14.21)$$

Clearly A and B should be of the lowest resistance that can
be achieved with adequate resolution of B and without overloading the bridge supply. The main difficulties in the
assembly of the bridge network is maintenance of purity of
the bridge elements and the elimination of significant phase
delays in connecting leads (given by $\omega/\sqrt{(LC)}$). These
difficulties can be minimised if all resistors are fixed, of

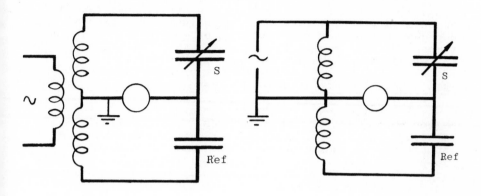

Fig. 14.12 Simple substitution   Fig. 14.13 Lynch bridge
          bridge

low value, and are physically small and all variable
elements are capacitances. A good solution to this problem
is shown in Fig. 14:11. The balance conditions are: (C
should be >10S).

$$\frac{j\omega X}{1 + j\omega X (R_x + r)} = \frac{j\omega S}{1 + \omega^2 SCrR + j\omega S (r + R + \frac{GR}{S})} \quad (14.22)$$

if tanδ <0.1 and r$\sim$1Ω then ω^2CSRr <10^{-5} so the balance
simplifies to X = S and tanδ = ω(C + S)R.   (14.23)

14.5.3 Substitution Methods

If the precision required makes unacceptable the
uncertainties, from unknown stray capacitances or unknown
inductive reactances, then substitution methods must be
used. The classical circuit uses voltage ratio arms, one
reference-arm containing a fairly good capacitor and the
standard and unknown arm as shown in Fig. 14:12. The bridge
is balanced with the unknown X, in parallel with the
standard S and X is then removed and S is increased in value
to restore the balance conditions. A phase angle balance is
also required at each step, the circuit of Fig. 14:10 being
appropriate. The capacitance of the unknown is then the
change in capacitance of the standard and it is worth
observing that an additional advantage is that a variable
capacitor can be calibrated more accurately in changes in

capacitance than in total capacitance. The loss tangent must be computed from the change in conductance and the change in capacitance. Lynch (14.5) has shown that the classical network is not a true substitution method because the load on the voltage arm supplying the standard and un-known arm is not constant due to the stray capacitance associated with the unknown. The solution proposed by Lynch depends on the fact that the load on the reference arm does remain constant and so the generator is applied directly to the standard/unknown arm as shown in Fig. 14:13.

14.5.4 Measurements at Radio Frequencies

As the frequency of the measurement is increased a method making use of the properties of tuned circuits is employed. Note first that for any impure capacitance or inductance the loss tangent is the ratio of the active or "in-phase" current component Ip, to the reactive or "quad-rature" component Iq so that $\tan\delta = Ip/Iq$. Now consider a tuned circuit with a capacitance in parallel with an inductance, fed from a constant current constant frequency supply. Define a resonance condition as one in which the total quadrature current is zero so that

$$Iq = Iq_L + Iq_C = 0 \qquad (14.24)$$

so that the input current is purely active and has the value

$$Ip = Ip_L + I_{p_C} = Iq\ (\tan\delta_L + \tan\delta_C) = Iq(\tan\delta) \quad (14.25)$$

rewrite this as $Ip = V\omega C_o\ \tan\delta$

C_o is found to be the total parallel capacitance of the tuned circuit. Now place in parallel with this tuned circuit a new capacitance ΔC whose magnitude is such that it too takes a current equal to $V\omega C_o\ \tan\delta$. This additional current is of course purely reactive and because the total input current is fixed the voltage across the total parallel combination falls from V to $V/\sqrt{2}$.

Hence if the capacitance of the original circuit is increased from the resonant condition by ΔC to reduce the voltage developed across the circuit to a value of $1/\sqrt{2}$ of the original we can write: $V\omega\Delta C = V\omega C_o\ \tan\delta$ from which $\tan\delta = \Delta C/Co.$ (14.26)

This is an important result because the expression for $\tan\delta$ is independent of frequency and is made up of two values ΔC and C_o that can be evaluated to a high precision. Even greater precision is obtained if the variable capacit-ance is changed by $-\Delta C_1$ and $+\Delta C_2$ to give two readings on either side of resonance when:

$$\tan\delta = (C_2 - C_1)/2C_o \qquad (14.27)$$

These properties were used to develop a means of measurement of loss tangent by Hartshorn and Ward in 1936 since when there has been a great advance in radio frequency measurements so that many of the difficulties described in the original paper have disappeared. Indeed, the test can now be made quite simply with a modern Q meter.

The precision available from the method depends in part on the quality of the inductances used to obtain resonances. At the high frequency end of the range (say from 1MHz upwards) this is not a limiting feature but at frequencies below 100kHz the Q of coils of acceptable size is seldom adequate.

In the method described the voltage across the coil is varied as the circuit is detuned on either side of resonance and it is very important that the inductance be strictly linear. The coils must therefore be air-cored and the high values of Q obtained with ferrite-cored coils cannot be exploited. This is unfortunate because when the method of measurement is changed from bridge to resonant circuits the highest possible precision is required near to the point of change otherwise a large overlap in frequency is required to eliminate the risk of spurious deviations from the spectral characteristics. Ferrite cored coils can be used, however, if the voltage and frequency are constant for any one measurement, and the circuit of Fig. 14:14 was developed to achieve this. The signal voltage, high enough to be rectified efficiently is attenuated by a capacitance potential divider having the tuned circuit connected in series across the lower element. The potential divider is adjusted until the reduction in applied voltage just compensates for the circuit magnification. Equal peak

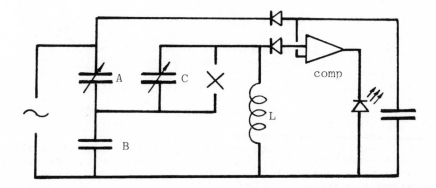

Fig. 14.14 Low r.f. reciprocal Q meter

voltages are then developed by the two rectifier circuits as demonstrated by the comparator to which the two circuits are applied.

At resonance the series tuned circuit presents a resistive load of value $r^1 = (\tan\delta)/\omega C$ across the output of the capacitor potential divider. The voltage injected into the tuned circuit is therefore:

$$\nu = V_{in} \frac{\{r^1/1 + j\omega Br^1)\}}{\dfrac{r^1}{1 + j\omega Br^1} + \dfrac{1}{j\omega A}} = \frac{V_{in} j\omega A \ (\tan\delta)/\omega C}{1 + j\omega(A + B)(\tan\delta)/\omega C}$$

(14.28)

$$\text{output voltage } V_o = \nu/\tan\delta = \frac{V_{in} jA}{C + j(A + B)\tan\delta} \qquad (14.29)$$

$$\text{for } |V_{in}| = |V_o|\tan\delta = \frac{\sqrt{A^2 + C^2}}{A + B} \qquad (14.30)$$

Note this is again independent of frequency and if A>10C and $\tan\delta$ <0.01 A can be calibrated directly in $\tan\delta$ to within 1%.

14.5.5 Measurements at Frequencies Below 50Hz

Although the gradual extension of the measurement technology to higher and higher frequencies has been entertaining it has not yielded very much additional information on the nature of dielectrics. The reason for this may be seen if the variation of permittivity and $\tan\delta$ with frequency is remembered. (See Chapter 5, Fig. 5.8). Few materials conform to the classical Debye curves but exhibit a more gradual change spread over many decades of frequency The frequency at which the slope $d\varepsilon_r/d(\ln\omega)$ is a maximum is the same frequency at which the value of $\tan\delta$ is a maximum. It is unlikely that a material will be used at power frequency if the frequency of maximum $\tan\delta$ is also close to 50Hz at the working temperature. Examination of the characteristics of materials in general use suggests that most such materials have the $\tan\delta$ peak at a frequency very much lower than 50Hz. Variations in the dielectric properties over the range of frequencies from 50Hz upwards are therefore minimal particularly when compared with the riches available at very low frequencies.

To explore this range measurements must extend by about 5 decades down from 50Hz that is to about 0.1mHz. Bridge methods are clearly of little use, several hours being required to balance a bridge at 1mHz. Another difficulty is the production of the driving waveforms at these low frequencies. Not surprisingly therefore, many workers have explored the possibility of applying a step voltage to the test specimen. The Fourier Transform d.c. Step Response technique has already been discussed in Chapter 5.

If the instantaneous values of current are recorded at intervals from the initial application, a Fourier transform can be used to extract the component of current in phase with the voltage at each frequency in the range $f = 1/2t$, to $1/2t_2$ where t_1 is the time required for a minimum of say five recorded currents and t_2 is the total duration of the step voltage. A typical frequency range is 100Hz down to 0.1Hz.

Hamon (14.6) has shown that a good approximation can be obtained for a particular frequency from a single instantaneous current reading. This can be explained as follows:

Suppose a square wave of voltage of frequency f be applied to the specimen. The current will give a peak value at the beginning at times equal to multiples of π and will be of the form: $i = (t)^{-1}$ during each half cycle as shown in Fig. 14:15.

The value of I' from which $\tan\delta$ may be derived, can be computed from the peak value of the fundamental component of the current in phase with the voltage. From Fourier this peak value is given by:

$$I_p = \frac{2}{\pi} \int_o^{\pi} \frac{\sin\omega t}{t} \quad t = \frac{2}{\pi} \int_o^{\pi} \frac{1}{t} \left(\frac{\omega^2 t^2}{2} + \frac{\omega^4 t^4}{4.1} + \frac{\omega^6 t^6}{6.1} + \right.$$

$$\left. \frac{8_t 8}{8.1} \quad \ldots \right) dt = 1.05 \tag{14.31}$$

This value of 1.05 must be multiplied by two because the initial application of a step function starting from zero will result in one half the current obtained when the voltage changes sign. It must be divided by $\pi/4$ to allow for the fact that the peak value of the fundamental sinewave component of a square wave has a peak value $4/\pi$ times the height of the square wave.

This gives $I^1_p = \frac{2.1\pi}{4}$

Now at some instant the instantaneous value of the current flowing into the sample will be equal in value to this value of $2.1\pi/4$. This coincidence occurs at

$\{ \frac{4}{2.1\pi \times 2\pi} \}$ {time for one cycle} $= 0.1tp$.

If the capacitance is measured at a frequency high enough for a bridge to be used a first approximation to $\tan\delta = Ip/V\omega C$.

An error has crept into this calculation because the value of C has been measured at the wrong frequency. However, a correction can be made for this error from the relation between $\tan\delta$ and change in permittivity (14.7) namely:

$$\frac{\Delta\varepsilon'p}{\varepsilon p} = \left(\frac{2}{\pi} \quad \ln \frac{\omega_2}{\omega_1} \right) \tan\delta \tag{14.32}$$

Hence, if values of Ip are obtained at intervals in
frequency less than one decade a step by step correction in
ε_p and tanδ can be made down to the lowest frequency.
    In this example the relation i = 1/t was assumed. The
original finding by Hopkinson was that the denominator
varied from about $t^{0.8}$ to $t^{1.3}$ for most materials and Hamon
showed in his paper that his approximation held, with an
acceptably low error over this range of powers of t.

14.6 PARTIAL DISCHARGE MEASUREMENTS

    When an insulation system is incorporated into a piece
of capital equipment it acquires two properties or measures
in addition to the four properties mentioned earlier,
namely discharge inception voltage (V_i) and discharge
magnitude normally quoted in pico Coulombs (pC). Curiously
the detection of partial discharge is very nearly as old as
the measurement of loss tangent, the simplest and first test
being made about 60 years ago. At that time the test was a
simple "yes/no" test, the discharge being detected
acoustically although the inception voltage and extinction
voltage were measured.
    Why was this hissing test so successful? Because it
exploited the natural geometry of the spout bushing to
amplify the sound output and to give a good signal to noise
ratio. When the hissing test was applied to through
bushings and to bushings of higher voltage, it became clear
that the method was no longer suitable and the search for
new methods started. There is, then, no best method of
discharge measurement, there are methods best suited to
particular test specimens characterised by material and
geometry, and the problem is to choose the most apposite
method and to apply this method in the most sympathetic
manner.
    It is now known that this listening test (known as the
"hissing" test) had a sensitivity of about 100pC on 33kV
bushings, and it is interesting to consider the performance
of these old bushings in the light of that test. Accoustic
detection is still used refined by the development of
contact transducers and operation at ultrasonic frequencies
where the signal to noise is much more favourable.

14.6.1 Survey of Methods

    When the self capacitance of a void or cavity in an
insulation system is discharged, there is a conversion of
electrical energy into heat, light and sound. It is
unlikely that a detector designed to be sensitive to the
heat available from a discharge will ever be competitive
with one designed to respond to light or sound. If a test
specimen may be stressed for a long period so that an
exposure time of several hours may be realised, an ultimate
sensitivity of a few picocoulombs may be expected for a
wide aperture camera and fast film. This may be the most
sensitive means of detecting discharges in a test specimen
which introduces a large attenuation between the discharging

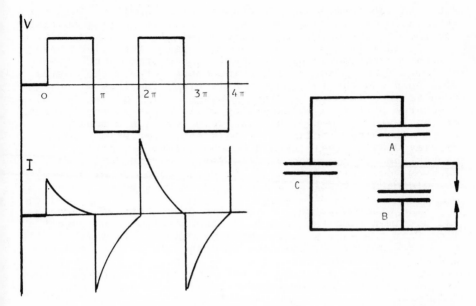

Fig. 14.15 Square voltage
wave and resulting
current wave.

Fig. 14.16 Discharging
void circuit.

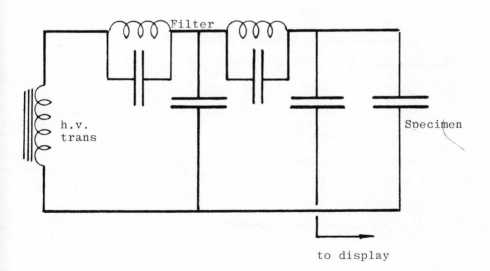

to display

Fig. 14.17 Mole input circuit

site and the terminals. An example of a sample of this sort
is a cable or alternator bar with a semi-conducting coating
applied for the purposes of stress relief.

In addition to this energy conversion there is a change
in the charge stored by the insulation system. One of two
extremes takes place. If the impedance between the source
of supply and the test specimen is high, the loss of charge
results in a loss of terminal voltage. With a symmetrical
discharge pattern a voltage step will appear at the terminal
of the test specimen each half cycle for each discharging
site. The high impedance for this characteristic to be
displayed effectively should be capacitive as in the
capacitance bridge type of detector. The impedance, may,
however, be a high resistance in which case the display of
the discharge signal is not so readily related to the actual
magnitudes of the discharge.

The other extreme is a low impedance between source
and specimen when the loss of charge results in a current
flow from the supply to restore the charge. In most
discharge detectors, this low impedance is a tuned circuit
and the step of current causes the circuit to ring. The
several types of detector in this category are characterised
by the bandwidth of the tuned circuit and the amplifier
following:

The restoration of charge from the supply contributes
to the dielectric loss of the specimen and any means of
measuring dielectric loss over a range of the voltages from
just below the discharge inception voltage to the highest
voltage of interest for the specimen is, in fact, a
discharge detector. It is seldom, however, that direct
measurement of watts is possible and a bridge method,
Schering bridge or inductively coupled dielectric loss
bridge, is usually required.

14.6.2 Description Of The More Common Methods

14.6.2.1 Use of dielectric loss bridge at 50Hz. At a
voltage below V_i the measured value of loss tangent
represents the dielectric loss in the solid insulation;
above V_i an additional contribution is made to the measured
value of loss tangent by the energy-loss due to the partial
discharge. For a single discharge site dissipating 1000pC
at 10kV rms in a sample of 1000pF, the increment of loss
tangent

$$\Delta\delta = \text{discharge energy/stored energy}$$
$$= (\tfrac{1}{2} \times 1000pC \times 10kV \times \sqrt{2})/(\tfrac{1}{2} \times 1000pF \times (10kV)^2)$$
$$= 0.707 \times 10^{-4} \tag{14.32}$$

This increment is just about the minimum detectable
signal. The calculation assumes that the discharge occurs
close to the voltage-peak; for discharges occurring before
the peak, the increment of loss tangent is lower.

14.6.2.2 The dielectric loss analyser (D.L.A.) In order to
add the separate discharges taking place in one half cycle,
the increments (or decrements) of charge must be stored

during the counting time without loss. Hence the time
constant of the integrating capacitor must be large
compared with the time of half a cycle. The D.L.A. achieves
this desirable property by the complete absence of any
resistive path between the test specimen and the supply.

The analyser is basically a four capacitance bridge,
two of the capacitors serving mainly to provide a reference
potential against which changes in the terminal voltage of
the test specimen stores the change in charge of the
specimen and the resulting voltage change is displayed on a
sensitive C.R.O. The bridge is initially balanced at a
voltage high enough to give adequate sensitivity but low
enough to give a clear horizontal line at balance and with
an ellipse when slightly unbalanced. The voltage is then
raised to the required value and any discharge unbalances
the bridge to produce, typically, a parallelogram display
on the C.R.O. screen.

If the cavities in the test specimen are within a
narrow range of sizes in the direction of the field (as
would be expected with wrapped insulation) the sides of the
parallelogram are straight with two sides horizontal. If
the range of cavity sizes is large (say with moulded
insulation) the sloping sides of the display curve slightly.
Similarly if the solid dielectric loss in the material has
a non-linear relation with the voltage the horizontal sides
of the display are found to curve.

From the conditions of balance the capacitance and
solid dielectric loss can be deduced; from the height of the
loop trace display the total charge transfer can be
calculated. The area of the trace is proportional to the
discharge energy and the ratio of (inception voltage)/
(applied voltage) is equal to the ratio of (horizontal
component)/(trace width). In addition to this quantitative
information the presence of any curvature can be inter-
preted as mentioned earlier. The lowest discharge
detectable is a function of voltage range and specimen
capacitance and in the commercially available model is at
best about a few hundred thousand picocoulomb in the sample
described in 14.6.2.1. This sensitivity may sound to be
poor but it should be remembered that the instrument is an
integrating device and this minimum value may be composed
of several smaller discharges. In fact, for most measurements it is quite adequate.

It is possible to interpose an amplifier between the
bridge output and the C.R.O. but the demands on the
amplifier are severe and a gain of ten or twenty is all
that should be aimed for. With a gain of ten times or more,
it is generally necessary to improve the waveform of the
applied voltage. This is not particularly difficult and a
tuned circuit technique can usually be applied.

14.6.2.3 <u>Direct extraction of the discharge signal</u> When
the equivalent circuit of a discharging insulator (Fig.
14.16) is examined it is clear that a direct electrical
signal can be detected at the terminals of the specimen, a
fact exploited by most methods employed at present to

measure partial discharge. In Fig. 14:16 the capacitance of the discharging cavity is B, C is the parallel capacitance presented by the rest of the specimen to the series combination of B and A, the capacitance of the cavity to the electrodes. Let V_o be the instantaneous applied voltage just before breakdown of B; at this moment the voltage across B is $V_B = V_oA/(A + B)$.   (14.33)

The charge on B, $Q_B = V_B.B = V_oAB/(A + B)$ (14.34)

The breakdown of the cavity can be represented by a step voltage $-V_B$ impressed on the cavity. This step will appear at the terminals with the attenuated value,

$$V_o = V_BA/(A + C) = V_oA^2/(A + B) (A + C)   (14.35)$$

The capacitance of the system when B is short-circuited by the breakdown path is (C + A) so that the apparent loss of charge by the specimen, observed at the terminals is
$\Delta Q = V_o (C + A) = V_oA^2/(A + B)$ (14.36)

The true charge transfer is $Q_B + \Delta Q_A$

$$= V_o \frac{AB}{A + B} + \frac{A^2C}{(A + B((A + C)} = V_oA \text{ very nearly since } C>>A$$
    (14.37)

The difference between apparent and true charge transfer is the factor $\frac{A}{A + B}$ which typically has the value ~ 0.1. There is no way of evaluating this factor other than very roughly. However, the apparent energy loss is found to be equal to the true value for:

apparent energy loss $= V_o Q = A^2V_o^2/2(A + B)$ (14.38)

and true energy loss $= \frac{1}{2}Q_BV_B = \frac{1}{2}V_oA V_oA/(A + B) =$

$$= A^2V_o^2/2 (A + B)   (14.39)$$

Whatever measure of the partial discharge activity is of interest, the step voltage at the terminals of the specimen must be measured. The first requirement therefore is the provision of a suitable impedance across which the voltage can be developed. In particular the specimen must not be short circuited at high frequencies by the output impedance of the high voltage supply. This supply must not introduce noise in excess of some tens of microvolts if reasonable sensitivity is to be maintained.

Two approaches to the measurement of this step voltage have been developed. The most significant step forward was furnished by Mole (14.8) introducing the first of a series of tuned circuit discharge detectors with the basic circuit of Fig. 14:17.

The test specimen, the high voltage discharge-free coupling capacitor and an inductance form a tuned circuit which is caused to ring by the step voltage resulting from the discharge. If the detector has a narrow bandwidth the ringing continues for many cycles and the output from the high frequency amplifier has a rise time greater than one cycle so that the highest peak in the train is not the first half cycle. Clearly then the sensitivity of the system is a function of the bandwidth among otherparameters and tends to rise as the bandwidth decreases. This is also true of the signal to noise ratio in an actual experimental location. Narrow band detectors, however, have the disadvantage that if the discharges are not resolved, unpredictable errors can arise due to the partial super-position of responses of more than one discharge. The narrower the pass band the fewer discharges per cycle that can be resolved.

As the bandwidth is increased the height of the first peak increases until when the bandwidth exceeds the centre frequency the first peak is the highest, resolution is a maximum and the errors due to superposition are minimised. The general rule is that a wide band detector is necessary to resolve many discharges and to examine the discharge pulse wave-shape but that a narrow band detector may be necessary to obtain the required sensitivity under noisy conditions.

It is usual for tuned circuit detectors to incorporate a C.R. tube display although with narrow band detectors the output from the high frequency amplifier is usually recti-fied and smoothed so that the display shows the envelope rather than the individual cycles of oscillation. This is done to improve the clarity of the display and improves amplifier stability.

Rather than restrict the bandwidth unduly it is better to use a modern swept frequency spectral analyser as the detector. This automatically exploits any "windows" in the noise spectrum to yield a high signal to noise ratio.

The second approach is an extension of the measurement of radio noise from high voltage apparatus which, for historical reasons followed a separate development from that of discharge measurement even though the two phenomena have much in common. Not unnaturally when the two techniques merged the use of standard radio noise meters were adapted to the measurement of discharge. The advantages claimed for the system are that the use of familiar equipment should lead to greater reliability, and there is a large body of experts in the use of radio noise meters, and that the recording of meter readings is easier than the recording of oscilloscope traces. This latter claim is best defended when test programmes are being automated.

A coupling capacitor and a resistor form a simple high pass input circuit to the amplifier which normally has a narrow bandwidth with a centre frequency of typically 1MHz. The output from the amplifier can have a short time constant and peak reading, a long time constant and integrate the individual discharges or, as is usual, a short charging time

constant and a long discharging time constant to give a
"quasi-peak" reading.

14.6.2.4 Calibration   A periodic and tuned detectors in
which the response is shifted in real time require the
injection of a calibrating signal. This signal must be
generated in a manner which permits ready calculation of
the resulting pulse height and must reproduce the pulse
shape of a discharge as nearly as possible. It is injected
into the test circuit through a small capacitor and the
value of this capacitor must be accurately known. For
example, its value must be large compared with the stray
capacitance introduced by connections.
    Apart from the calibration of the detector it is often
necessary to calibrate the high voltage test circuit to
take stray capacitances into account.
    Even more pronounced is the need for calibration when
the spectrum analyser is used. If the chosen frequency is
very high, say, above 100MHz the rise time of the pulse
arriving at the terminals of the test specimen from the
partial discharge site may be so long relative to one
quarter cycle of the detection frequency that the signal
voltage is significantly reduced. A measurement made at
this high frequency appears therefore to under estimate the
magnitude of the discharge. A test specimen in which this
effect is aggravated is one that is physically long - a
cable or switchboard for example. In addition to the
attenuation of the discharge pulse from the discharge site
to the terminals, there is often a distortion of the
pulse front which increases the rise time.
    It is clear from the foregoing that the calibrator
must reproduce as accurately as possible the rise-time
of the discharge itself.

14.6.2.5 Interpretation of partial discharge results
The usefulness of partial discharge measurements is
enhanced if the values can be related to remaining life.
The best progress in estimating a relation can be made
where many specimens, nominally identical, are in
service and a large sample has been tested. In this
instance the specimens can be ranked in order of discharge magnitude and if the annual failure rate is known
the fraction of the sample expected to fail within one
year can be identified. Unfortunately one characteristic
of partial discharges, particularly in the last twelve
months of life, confuses this simple approach. It is
found that the more energetic discharges do not occur
continuously but exhibit periods of inactivity. Measurements made on installed plant, therefore, may not be made
at a time coinciding with the incidence of the more
energetic discharges. Such plant therefore is placed too
low in the ranking order.
    This effect can be minimised if the plant can be dis-
connected from the supply, excited from a separate test
transformer and measurements made over a range of voltages

from V_i to an overvoltage of about 15%. Of course, this is a more expensive measurement than one made with the plant in service.

A characteristic feature of any statistical approach is that the specimens of interest lie on the tail of the distribution. The average value is of little interest but the maximum expected value is of paramount interest. Whatever the probability distribution the average value is always easy to determine but to estimate the value with a probability of occurring of, say, 0.01 requires a precise knowledge of the type of distribution.

It is becoming increasingly apparent that in the field of insulation measurement the distribution most seldom encountered is the Gaussian or 'normal' distribution. Why is this?

When a discharge magnitude is measured on an item of plant, the recorded value is the first of many discharge sites to burst into activity for a measurement at V_i, at a voltage above V_i the recorded value is the largest observed magnitude. In either instance the measurement is of a form of extremal value, akin to the strength of a chain. Gumbel has shown that extremal values fall into one of four main distributions and measurements of discharge magnitude on many items of plant have been found to conform to the Gumbel No. 1 distribution very closely as Fig. 14.18 confirms.

The range of values obtained on plant in service can range over five decades from 0.01 to 0.99 probability. However, this range covers fairly new plant to plant approaching the end of its life. An estimate of the range of discharge magnitude at working voltage can be made for some insulation systems known to have a maximum permitted value of about 100pC when new. The lowest value, other than zero, can be estimated from the combination of field strengths in the cavity and the solid insulation constructed from Paschen's curve for the breakdown of air, and is about 0.1pC.

All the measurements on a particular type of plant covering the full range of life from new to near failure can be fitted to a pattern of a Gumbel distribution from 0.01 to 0.99 probability moving upwards during the service life from discharge magnitudes ranging from 0.1pC to 100pC when new to a range 2pC to 2000pC when the pattern appears to fail. This suggests that aging is taking place during service at a rate dependant on discharge magnitude up to a magnitude of about 2000pC when the rate of deterioration appears to increase sharply.

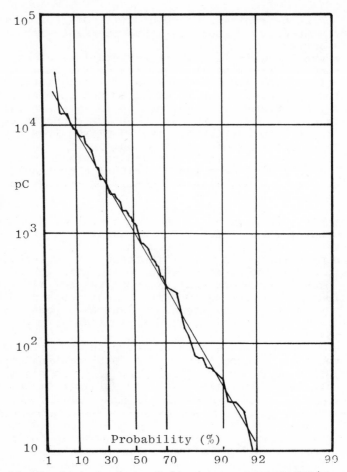

Fig. 14.18   Gumbel plot of discharge measurements

14.7 REFERENCES

14.1 Martin, R.G. and Patterson, E.A., 1953 <u>Proc. I.E.E.</u>
     <u>100</u>(11A) 68

14.2 Blumlein, A., 1928 Patent No. 323037

14.3 Glynne, A., 1952 <u>Bull. Elec. Eng.</u> Educ. <u>8</u> 69

14.4 Starling, S.G., 1941 (Book) Electricity and
     Magnetism (Longmans).

14.5 Lynch, A.G., 1956 <u>Proc. I.E.E.</u> <u>104(B)</u> 363

14.6 Hamon, B.V., 1952 <u>Proc. I.E.E.</u> <u>9</u> 9 151-155

14.7 Lynch, A.G., 1971 <u>Proc. I.E.E.</u> <u>118</u>, 244-246

14.8 Mole, G., 1953 <u>Proc. I.E.E.</u> <u>100</u>, 276

Chapter 15

Breakdown testing and measurements on installed equipment

A. W. Stannett

15.1 INTRODUCTION

Several kinds of schemes are available for the electrical evaluation of solids and composites, to meet various linked objectives. These are:-
(1) to determine if, and demonstrate that, the insulating system is basically capable of meeting the design requirement. These are called development, design and type tests and fall into two groups.
(a) rapid tests to determine definitive properties, and   (b) long-term tests to check ageing characteristics and long-term reliability.
(2) Quality control or quality assurance tests. These are rapid tests usually derived from those described in l(a) whose purpose is to ensure that the product being delivered against an order is similar to that which had been successfully subjected to type test and approved as suitable for the customers' needs.
(3) Condition monitoring and technical auditing. These are respectively "on" and "off-line" measurements aimed at determining the condition of insulating systems in service.
It is evident from the foregoing that measurements on insulating materials are inseparable from measurements on plant and that the interest is in determining not some fundamental property of the insulating system which is relevant to every application but rather how a particular system is chosen for and reacts to, the conditions that the design and usage impose.
The first step therefore is to understand what are these conditions and what are the sensitive properties of the insulating systems in the particular case. Often this understanding comes from the development tests of category 1. On this basis, each case must be considered on its own merits and it is proposed to discuss a few particular examples to illustrate the approach rather than to attempt to cover everything.

15.2   OUTDOOR INSULATORS

15.2.1   General Requirements.

Most overhead line transmission and distribution
systems are insulated with porcelain or glass suspension,
tension or post insulators.   Under certain conditions and
in particular places, composite insulators made from glass
fibre reinforced strength members with organic resin binder
and housings are attractive and are used.
Put at its simplest, the primary requirement for an
overhead line insulator is that it should support the line
mechanically under all foreseeable conditions.   This means
that if the voltage on the insulator exceeds its withstand
capability then, it should flashover without causing damage,
rather than puncture and risk losing mechanical strength.
In U.K. insulators are purchased to BS 137 (15.1)
which has a type test requirement which demonstrates their
dry lightning impulse strength, their wet switching impulse
strength and their wet power frequency strength.   There is
also a "sample test" requirement where a proportion of the
production batch of insulators is taken and tested virtually
to destruction in the temperature cycle test;   electromechanical failing load;   mechanical failing load; thermal
shock (for toughened glass insulators) power-frequency
puncture, porosity and a check on the galvanising.   Routine
tests are carried out on every insulator supplied and these
are a routine mechanical test to about half the failing load
(depending on the type), an electrical test to demonstrate
that the insulator flashes over rather than punctures and
for glass insulators only, a thermal shock test.

15.2.2   Pollution of Inorganic Insulators.

Insulators have to work under a very wide range of
conditions from desert to salt fog and important though the
tests are in the British Standard, it is the shape of the
insulator which governs how it will perform under polluted
conditions which has proved to be so important.   In the
U.K. the early insulator pollution work was done at Croydon
(15.2,15.3) but as industrial pollution became less
important relative to salt fog marine pollution, the
Brighton Insulator Testing Station was set up and is now a
sophisticated tool for assessing the pollution performance
of insulators up to 900 kV to earth (15.4-15.8).   The
simplest methods of comparing insulator performance are:-
   (a) the exposure time to flashover or frequency of
       flashover
   (b) the magnitude of the greatest current surge
   (c) the rate of counting at different current levels,
       e.g. 25, 150 and 250 mA equivalent RMS using
       electromagnetic counters.
A more sophisticated method shown in Fig. 15.1 and used in
UK utilizes a number of special explosive fuses designed to
disconnect when a flashover occurs thereby increasing the
lengths of the insulator (15.6-15.8).   The circuit breaker

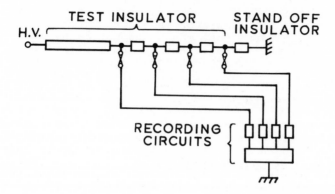

Fig. 15.1 Testing circuit for 1500 kV
          insulators at Brighton (Ref. 15.5)

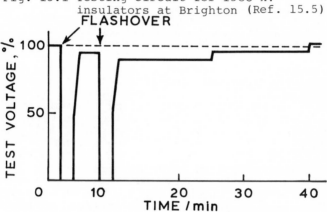

Fig. 15.2 Test voltage variations in
          response to flashovers (Ref. 15.5)

controlling the supply will also trip on flashover.   As
shown in Fig. 15.2, the regulator is arranged to run down
to half setting and when the breaker closes, the voltage is
increased steadily to 95% of the original.   If no flashover
occurs, it is increased after 15 minutes to a value slightly in excess of the original one and so on.   By using two
insulators with known performance as 'controls' it is
possible always to compare insulators one with another.

     This sort of approach is very time-consuming and is
supplemented, if not replaced, by artificial pollution tests
(15.9-15.11).   IEC 507 (15.12) describes two categories of
tests which are now commonly used.   In the first method the
insulator is set up in a chamber in which a fog of salt
water with variable salinity is produced.   The effectiveness of the insulator is judged by the salinity of the
solution it will support at its working voltage without
flashing over or by the voltage it will stand for a given
salinity.   Alternatively a solid layer of pollution can be
applied to the insulator the recipe being adjusted to
produce particular conductivities when the layer is wetted

by a clean water fog.   As with the salt-fog test the
effectiveness of the insulator is judged by the conductivity
of the pollution layer that can be supported by the
insulator or by the voltage that can be applied with a
given contaminant.   Both techniques are also used in withstand modes.

15.2.3   Composite Insulators.

Composite insulators consist of at least two parts, a
mechanically strong core, usually made of resin impregnated
glass fibre encased in a housing bearing sheds to provide
the necessary creepage distance (15.3, 15.7).   The housing
may be made from a variety of organic elastomers or resins.
Composite insulators have to withstand the same
mechanical and electrical stresses as any other insulator
and therefore need to pass much the same range of short-
term tests as insulators made of porcelain or glass.   They
differ in the sense that organic materials age and degrade
in service and the capability of withstanding degradation
sufficiently to give the required life needs to be assessed.
The subject has been discussed in some detail by Cojan et al
(15.14) from the point of view of insulators used on
transmission systems (and an IEC document detailing test
requirements is being produced) and by Bradwell and
Wheeler (15.15) from the railway application viewpoint.   The
European transmission experience reported, highlights the
vital importance of how the design deals with the inevitable
interfaces which occurs between the various components of
the insulator as they tend to be both mechanically and
electrically weak.   The most important is the end to end
interface between the strength member and the housing.   It
is also necessary to check the effectiveness of the bond
between individual glass fibres and resin in the strength
member by means of a dye penetrant and electric strength
test.   Composite insulators are more commonly used on the
railway systems (15.15) where their low mass, high strength
and toughness has permitted novel and economic designs to
be used with advantage particularly where space is
restricted such as under bridges.
In both these papers the importance of natural
exposure testing is demonstrated permitting attempts to be
made to simulate by accelerated tests the failure processes
observed.
It is, of course, a fundamental requirement that the
housing material does not "track" or "erode" severely when
subjected to wet sparking conditions whether the insulators
be for use on a 400 kV transmission line, a 25 kV railway
system or in an 11 kV switch chamber in a damp dirty
situation.

15.2.4   Tracking and Erosion Tests.

15.2.4.1 On complete insulators. A track is defined as a
permanent conducting path formed on the surface of an
insulator.   In this case we are concerned with carbonaceous

paths formed from the insulating material itself by the action of sparking under wet polluted conditions (15.16). It will be recalled that when current flows in a wet polluted layer on the surface of insulators, the heating effect causes some evaporation of water which leads to ever greater heating in those areas of lower conductivity. The result is local instability and the creation of a "dry band" across which, all of the voltage applied to the specimen appears (15.17). During the process of formation of a dry band and its subsequent interruption by droplets or trickles of water, the arcs which occur when the leakage current is interrupted, heat the insulator locally for brief times. This is often sufficient to induce chemical change in organic materials. These will either result in the production of a carbon-rich char (tracking) or loss of material (erosion) either as the carbon and hydrogen react with oxygen from the air to produce carbon oxides or the material simply depolymerizes (15.16; 15.18). The formation of a dry band is the first step in the flashover process (15.17) of an insulator (Fig. 15.3) and its position is governed inter alia by the shape of the insulator and the distribution of pollution on its surface. It therefore follows that development and type tests for tracking on insulators should be done on insulators of the right size and shape. This is often achieved by testing the insulator (or part of it) in a salt fog chamber for periods of the

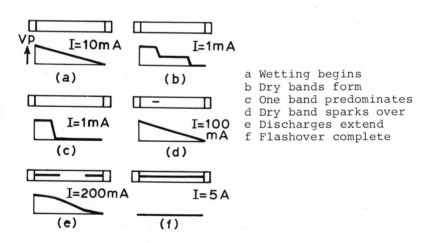

a Wetting begins
b Dry bands form
c One band predominates
d Dry band sparks over
e Discharges extend
f Flashover complete

Fig. 15.3 Typical voltage distributions
on a polluted strip (Ref. 15.5)

order of 1000 hours.   In the case of insulators for use on
the railway system in tunnels it is suggested that the
insulator be coated with a layer of solid pollutant containing iron oxide made conducting with a fog of de-ionised
water.   The iron oxide which in practice comes from the
brakes of the trains and accumulates where the insulator is
not washed by rain causes certain organic materials to fail
by burning under wet sparking conditions.

15.2.4.2 On materials. It is, of course, impractical for
manufacturers of materials to develop them using complete
insulators as test specimens and many different kinds of
tracking tests exist to compare and evaluate materials in
sheet or rod form.
    The differential wet tracking test (15.19) and the
dust/fog test (15.20) have served well and although they
have their attractions are less used now than the drop
tracking test (15.21) and the inclined plane test (15.22)
which are the only ones graced with IEC status.   In the
former, drops of a solution of ammonium chloride are
arranged to fall on the surface of the test sample to wet
the space between a pair of chisel electrodes between which
an adjustable test voltage is maintained (Fig. 15.4).
Basically, the voltage is determined at which a track is
formed on the surface after 50 drops of electrolyte have
been applied.   This voltage is the comparative tracking
index for the material and is a good indication of the
track resistance of materials which track fairly readily
but are nevertheless quite adequate for use at domestic
voltages.   The inclined plane test is much more severe and
is applicable to track resistant materials such as might be
contenders for housings for outdoor insulators.   In this
test, a film of electrolyte is arranged to trickle down the
back surface of a sheet (Fig. 15.5) and unless the test is
being used in a "withstand" mode, samples are rated in terms
of the voltage which causes a track to form in one hour.
There are many debates about the validity and precision of
these tests (15.23; 15.24).   For instance, there is little
doubt that the rate and type of the chemical reactions

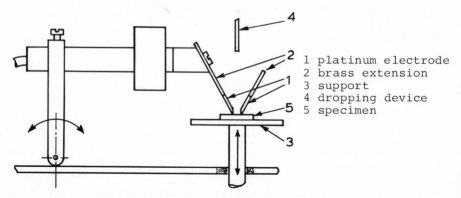

1 platinum electrode
2 brass extension
3 support
4 dropping device
5 specimen

Fig. 15.4 Drop Tracking Test

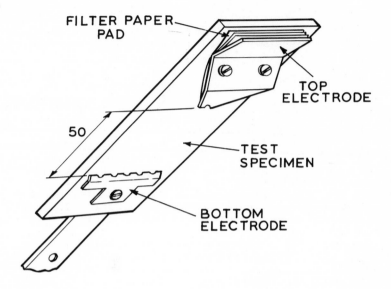

FILTER PAPER PAD

50

TOP ELECTRODE

TEST SPECIMEN

BOTTOM ELECTRODE

Fig. 15.5 Inclined Plane Arrangement

which lead to tracking or erosion depend on the energy
density during sparking. The energy density in the inclined
plane test for instance increases as the test voltage
increases and therefore this test would probably not simply
accelerate the rate of a particular type of breakdown to
produce the same classification of different materials as
would be achieved at a fixed voltage for very long times.
This is why some still like the old dust and fog test or the
Scandinavian wheel test (15.25) making a virtue of the
month-long testing times.   In the latter, the specimens
which are mounted on a wheel rotating at about 1 R.P.M. dip
into a solution of electrolyte and dry by ultra-violet lamp
during half of each revolution with the test voltage applied.
A salt-fog test is also sometimes used.   There are several
variants of detail but, (15.26-15.28) in principle, the
specimens, usually rods 6" x 1" dia. (like those used in the
wheel test) are aged in a salt fog with voltages of the
order of 10-13 kV applied.   From time to time, the voltage
is increased to flashover and the reduction in flashover
voltage caused by ageing used as a measure of resistance to
ageing.
        Most of these tests attempt to accelerate the effects
of contamination and dry band formation.   Except insofar
that the light emitted by sparking causes surface degradation
these tests do not simulate weathering caused by sunlight and

the weather, let alone accelerate these effects. It is, of course, common observation that the surfaces of synthetic resins craze and chalk in time when exposed to the sun and weather (15.29). This causes the surface to become more easily wetted by rain and become more conducting. It is for this reason that EdF devised a test (15.30) which incorporates weatherometer ageing as part of the test requirement.

15.3 POWER CABLES

Different types of cable are used to cover the range of say 11 to 400 kV. Up to 11 kV, extruded cables using PVC, XLPE or EPR are tending to be used, certainly in dry situations whilst oil and paper cables with metal sheaths are still favoured for the higher distribution voltages in the U.K. Elsewhere extruded cables are in more general use in spite of the water-tree problems. At the highest voltages (275 kV and above) fluid oil impregnated paper (OIP) is almost universally used as insulant either with a lead or aluminium sheath or installed in a pipe. Extruded cables are gradually creeping up the voltage rating scale but until design stresses can match those of O.I.P. the chances of cost parity between extruded and taped cables are slender.

Whilst the detailed test schedules vary from size to size and type to type there is a general pattern of test requirement (15.31-15.33). Cables with their associated terminations and joints are generally quite thoroughly type-tested mostly with a series of heat cycle tests; frequently 20 during which 1-1/3 x w.v. is applied to the cable whilst the conductor is heated to the maximum design temperature in 8-hour heating and 16-hour cooling cycles. During such an ageing test, the stability of the insulation is examined hot and cold daily usually by measuring the loss tangent/voltage characteristic. The impulse strength is measured after such an ageing test. The impulse strength is also measured after a bending test simulating the bending the cable undergoes during manufacture and installation. The impulse test requirement is a real one in the sense that a cable system will experience lightning and switching surges controlled by arc gaps or surge arresters, to the designed insulation co-ordination level. The heat cycle test is important in OIP cables in demonstrating long-term stability in the sense of freedom for oil expansion and contraction and shorter-term freedom from thermal runaway.

The case of extruded cable differs (15.34) mainly because the insulants are relatively sensitive to breakdown by partial discharge. Since there is no liquid insulating component impregnating and filling the spaces, they must be made virtually perfectly with no voids large enough to support discharge at the working stress or particulate contamination to induce tree breakdown anywhere in the whole length of the system. This is a most exacting requirement which has involved the development of extruded partially-conducting screens at conductor and sheath and means for extruding them near perfectly.   Since these components have different co-efficients of thermal expansion to the dielectric wall, the purpose of the heat cycle test is to demonstrate that the joins are sound and the separate layers do not part company during life.   Sensitive discharge tests usually replace the loss tangent/voltage test because the loss tangent of XLPE is very low.   Either the test is carried on longer than that for OIP or testing is done at a higher frequency, frequently 500 Hz to accelerate any discharge damage that might be occurring. Such is the need to demonstrate perfection in manufacture and installation.

At around 11 kV in power stations, the reliability of the generating set, which may be a nuclear one, depends amongst other things on the integrity of the cables connecting pumps and fans.   Consequently discharge tests are carried out as a routine on each drum length of cable delivered to site.   Further discharge tests are carried out in each installed length (15.34, 15.35) to ensure that the cable has not been damaged during installation and the terminations which are often the push-on type are properly fitted.   The costs of a failure in service and an outage far exceed the cost of these extra tests.

The discharge tests on installed cable can be repeated for off-line technical auditing if required but generally the only real worry with properly made cables, carefully tested as described, is water treeing (15.36, 15.37) caused by water penetration of damaged oversheaths or into the conductor due to faulty seals.   Claims have been made in USA (15.38) and Japan (15.39) that the presence of water trees can be detected by loss tangent and dielectric absorption tests but the author and his colleagues cannot support this claim.   They only know of the relatively clumsy techniques of checking oversheath integrity by using a "Megger" test between armour and earth and by checking the relative humidity of gas passed through the conductor space.   Hopefully something more elegant will appear in due course.

Cable joints and terminations can be monitored continuously in service for the development of discharges by using specially designed detectors.   These have found application during the early field trials of 132 kV extruded cable systems in U.K. although the technique was developed especially to monitor 275 and 400 kV OIP joints (15.40).

15.3.1 Tests on Cable Materials.

The foregoing section indicates that extruded insulation is gradually replacing lapped paper for the range 11-400 kV as we know it.   Lapped polythene/gas (15.32) and polypropylene/paper laminate/oil (15.33) insulation are possible future insulants particularly if we ever need to make cables for voltages much in excess of 400 kV.   The problems with extruded cables are the lack of resistance of the polyolefins to "treeing" (15.41) or channel propagation breakdown and to "water treeing" (15.42).   Very many papers

have been written on these two subjects, but there is no
answer yet.   The various excellent review papers describe
various tests for the resistance of material to these two
types of breakdown.   They are not standardised because
their relation to service needs has not been demonstrated;
however, their use as material development tools may well
prove to be as useful as the developments in extrusion
technology and maintenance of cleanliness of the polymer.

15.4   CHEMICAL METHODS OF MONITORING INSULATION IN SERVICE

As a general rule, plant installed in power systems
is well designed, made and maintained.   But occasionally
things go wrong and it is useful to be able either to
monitor the condition of insulation continuously as it
operates or to determine its condition so that it can be
repaired if needed, and if not, hopefully to determine its
remanent life by a technical audit.   When faults do develop,
they are usually local and either caused by local overheating or local partial discharging or sparking.   Since
only small amounts of insulation are involved, alternating
voltage tests, such as loss tangent, which measure an
average electrical characteristic seldom reflect their
presence.   Commonly only discharge tests and direct voltage
tests are capable of detecting them.   However, imagine try-
ing to do a discharge test on a 400 kV transformer in situ;
the effort entailed in disconnecting it from the system;
the charging current and size of test transformer and
blocking capacitor required and the cost of the loss of
supply capability.   In cases like this other methods have
to be considered and it is often convenient and effective to
seek chemical products of local discharges or points of
overheating.

15.4.1 Gas-in-Oil Tests.

The idea of detecting the presence of defects by
chemical analysis is particularly attractive for oil
impregnated paper insulation such as transformers, both
power and instrument, bushings and even cable joints
because the gaseous products are trapped by the oil which
can be readily sampled.   The problem is in devising
sufficiently sensitive analytical equipment and in learning
how to use the results.   The gas chromatograph has been
sufficiently developed for the purpose and in the early
1960's the CEGB analysed the oil, from around 1000 large
transformers some of which contained faults (15.43).
Analysis of the results, bearing in mind Halstead's
theoretical work showed that nature of the different gases
indicated their origin, e.g. hydrogen and hydrocarbons from
the oil, carbon oxides from the paper and the type of
degrading process which produced them (15.44).   In
practice, the ratio of the concentration of four pairs of
gases is used as an indication of the energy of the process
including arcs and various ranges of thermal degradation.
This work has been confirmed by international collaboration

and is now incorporated in the Standards ANSI/IEEE C57.104 (1978) and IEC Publication 599 (15.45).

These techniques are sufficiently advanced now to be used on new transformers on test in the Works, yielding 'fingerprints' of their likely history.

Even more recently, techniques are being developed to detect the presence of soluble degradation products of overheated paper to supplement the detection of overheated paper by the presence of the carbon oxides. The chief diagnostic degradation product is furfural and it is detected by high performance liquid chromatography.

15.4.1.1 On-line monitoring. There are occasions, maybe when there is no laboratory facility near the transformer or there is a particular need to try to "nurse" a transformer containing a fault by controlling its loading when continuous on-load gas-in-oil analysis is helpful. A Canadian device based on a small fuel cell for monitoring hydrogen is available (15.46; 15.47), so too is a British device for monitoring the hydrocarbons (15.48).

15.4.2 Gas-in-Gas and Generator Condition Monitoring.

Electrical methods of determining the location of discharges in stator insulation with the machine out of service and of continuously monitoring for discharges and arcing are available (15.49). Even so, analysis of the hydrogen gas (and cooling air of some motors) are useful techniques now very generally applied for monitoring machines to detect the onset of overheating of the insulation for whatever reason (15.50).

"Conditions monitors", as they are called work on the principle that overheated organic insulation produces, not only gaseous fragments but submicron particles which are carried in the hydrogen coolant. The monitor contains an ionization chamber with a weak alpha radiation source arranged to generate an ion current flow across a stream of hydrogen drawn from and returned to the alternator. If this gas contains submicron particles produced by thermal decomposition, the current flow is reduced at a rate proportional to the concentration of the particles, the reduction being a measure of the overheating. This output current is continuously recorded and can be used to activate an alarm. Devices of this sort respond to oil mists but this can be overcome by using a heated chamber.

Variants are available in which the particles can be collected on a filter and identified by their fluorescence under U.V. irradiation (15.51).

Condition monitors indicate that insulation is overheating somewhere in the system. One may also be able to identify the nature of the material which is overheated which can be very helpful.

Methods are currently being developed and are in use on an exploratory basis to enable the general position of the overheating to be located. It is being done by the use of tagging compounds. These are compounds which are in-

corporated into paints applied to specific areas of the
machines, which when overheated can be detected in the
hydrogen and analysed by gas chromatograph thereby identify-
ing the general area in which the overheating is occurring
(15.51; 15.52).

15.5   CONCLUSION

IEC BSS and other standard test procedures are
essential aids to buying and selling materials and equipment.   These tests are often helpful in assessing
insulation and insulating systems but for design and
operational purposes each case needs to be taken on its
merit and the method of assessment suited to the engineering
need.

15.6   ACKNOWLEDGEMENT

The work was carried out at the Central Electricity
Research Laboratories and is published with the permission
of the Central Electricity Generating Board.
Material from BS 5604 and BS 5901 appear by
permission of the British Standards Institution,
2 Park Street, London W1A 2BS from whom complete copies can
be obtained.

REFERENCES

15.1   BSS 137 1982, Insulators of ceramic material or
glass for overhead lines with a nominal voltage
greater than 1000 V.   Part 1: Methods of test
Part 2: Requirements.

15.2   Forrest, J.S., 1942, Characteristics and performance
in service of H.V. porcelain insulators.   J.I.E.E.
89, II, 60.

15.3   Forrest, J.S., Lambeth, P.J., and Oakeshott, D.F.,
1960, Research on the performance of high voltage
insulators in polluted atmospheres, Proc.IEE, 107A,
172.

15.4   Lambeth, P.J.,Looms, J.S.T., Sforzini, M.,
Malaguti, C., Porcheron, Y. and Claverie, P., 1970,
International research on polluted insulators,
CIGRE Paper 33.02.

15.5   Lambeth, P.J., 1971, Effect of pollution on high
voltage outdoor insulators, Proc.IEE IEE Reviews,
118, 1107.

15.6   Lambeth, P.J., 1973, Insulators of 1000-1500 kV
systems, Phil.Trans.R.Soc., 275, 153.

15.7   Lambeth, P.J., Looms, J.S.T., Roberts, W.J. and
Drinkwater, B.J., 1974, Natural pollution testing of

insulators for UHV transmission systems, CIGRE Paper 33.12.

15.8   Houlgate, R.G., Lambeth, P.J. and Roberts, W.J., 1982, The performance of insulators at extra and ultra high voltage in a coastal environment. CIGRE Paper 33.01.

15.9   Ely, C.H.A. and Lambeth, P.J., 1964, Artificial pollution test for high voltage outdoor insulators, Proc.IEE, 111, 991.

15.10   Lambeth, P.J., Looms, J.S.T.,Leroy, G., Porcheron, Y. Carrara, R. and Sforzini, M., 1968, The salt fog artificial pollution test, CIGRE Paper 25.08.

15.11   Ely, C.H.A., Kingston, R.G. and Lambeth, P.J., 1971, Artificial and natural pollution tests on outdoor 400 kV substation insulators. Proc.IEE, 118, 99.

15.12   IEC Publication 507, 1975, Artificial pollution tests on high voltage insulators to be used on a.c. systems.

15.13   Stannett, A.W., Lambeth, P.J., Parr,D.J., Scarisbrick, R.M., Wilson, A. and Kingston, R.G., 1969, Resin-bonded glass fibre outdoor h.v. insulators. Proc.IEE, 116, 261.

15.14   Cojan, M., Perret, J., Malaguti, C., Nicolini, P., Looms, J.S.T. and Stannett, A.W., 1980, Polymeric Transmission Insulators: Their Application in France, Italy and the U.K.   CIGRE Paper 22.10.

15.15   Bradwell, A. and Wheeler, J.C.G., 1982, Evaluation of plastics insulators for use in British Railways 25 kV Overhead Line Electrification, Proc.IEE, 129B, 101.

15.16   Parr, D.J. and Scarisbrick, R.M., 1965, Performance of synthetic insulating materials under pollution conditions, Proc.IEE, 112, 1625.

15.17   Hampton, B.F., 1964, Flashover mechanism of polluted insulation, Proc.IEE, 111, 985.

15.18   Groves, D.J. and Kaye, P.H., 1979, Tracking events within a dry band.   DMMA IEE Publication No. 177, 270.

15.19   ASTM D 2302 - 64 T Test for differential wet tracking resistance of electrical insulating materials with controlled water-metal discharge.

15.20   ASTM D 2132-62T Dust and fog tracking and erosion resistance of electrical insulating materials.

15.21   IEC 112 BS 5901, Method of test for determining the comparative and proof tracking index of solid insulating materials under moist conditions.

15.22   IEC 587 and BS 5604.   Method of test for evaluating resistance to tracking and erosion of electrical insulating materials used under severe ambient conditions.

15.23   Mason, J.H. and Watson, J.F., 1979, Some factors which affect the tracking and erosion resistance of organic materials. DMMA IEE Publication No. 177, 278.

15.24   Weller, M.G., 1979, Variation in the results of inclined plane tracking tests: Some factors controlling the surface tracking activity. DMMA IEE Publication No. 177, 282.

15.25   Kurtz, Mo., 1971, Comparison of tracking test methods IEEE Trans. on Electrical Insulation EI6, 76: (see Appendix for the original wheel test which forms the basis for the modern Scandinavian development, in which the spray is replaced by an electrolyte bath).

15.26   Mullen, A. and Dakin, T.W., 1982, Wet tracking and erosion evaluation of non-ceramic insulation for outdoor use.   IEEE Conference Philadelphia, June 82 CH 1780-6-EI 238.

15.27   Jolly, D.C., A test method for determining the outdoor lifetimes of polymer overhead transmission line insulators.   ibid 248.

15.28   Reynaert, E.A., Orbeck, T. and Seifferly, J.A., Evaluation of polymer systems for outdoor HV insulator application by salt fog chamber ibid 242.

15.29   Tourneil, C. de Natural and accelerated climatic ageing of polymer materials for HV insulation. ibid 233.

15.30   Folie, M., Perret,J. and Fournie, R., 1979, Resistance to tracking of synthetic insulating materials subjected to a climatic ageing. DMMA IEE Publication 177, 274.   (Electricite de France specification HN 26-2-20 - Tenue au cheminement des materiaux isolants synthetiques soumis à un vieillissement climatique.)

15.31   Stannett, A.W. and Gibbons, J.A.M., 1976, Case history of the development of the design concepts for a novel cable.   Electronics and Power, 22, 299.

15.32 Gibbons, J.A.M. and Stannett, A.W., 1973, Present state of development of a gas-pressurized lapped polythene cable for e.h.v. transmission, Proc.IEE, 120, 433.

15.33 Arkell, C.A., Edwards, D.R., Skipper, D.J. and Stannett, A.W., 1980, Development of polypropylene/ paper laminate (PPL) oil-filled cable for U.H.V. systems. CIGRE Paper 21.04.

15.34 GDCD Standard 17 (formerly CEGB Standard 095101) 6350/11000 volts extruded solid insulation cables associated with GDCD Standard 21 (formerly CEGB Standard 099905) Cable Insulations having reduced fire propagation.

15.35 Wilson, A., 1974, Discharge detection under noisy conditions. Proc.IEE, 121, 993.

15.36 Fournie, E. and Auclair,H., 1979, Effect of water on electrical properties of extruded synthetic insulation application cables. IEEE Paper A79 411-O. (Read at PES Summer Meeting, July 1979.)

15.37 Lyle, R. and Kirkland, J.W., 1981, An accelerated life test for evaluating power cable insulation. IEEE Paper 81 WM 115-5 (read at PES Winter Meeting February 1981).

15.38 Bahder, G., Eager, G.S., Suarez, R., Chalmers, S.M., Jones, W.H., Mangrim, W.H., 1977, In-service evaluation of P.E. and XLPE insulated power cables rated 15-35 kV. IEEE Trans. P.A.S. 96 1754.

15.39 Issiki, S. and Yamomoto, M., 1979, Development of water tree detecting device. Fujikura Technical Review, 20.

15.40 Wilson, A., Nye, A.E.T. and Hopgood, D.J., 1982, On-line detection of partial discharges in H.V. plant. BEAMA 4th Electrical Insulation Conference, 233.

15.41 Mason, J.H., 1978, Discharges. IEEE Trans, E.I., 13, 211.

15.42 Nanes, S.L. and Shaw, M.T., 1980, Water treeing in polythene - a Review of Mechanisms. IEEE Trans. on E.I. EI-15, 437.

15.43 Rogers, R.R., 1975, U.K. experience in the inter-pretation of incipient faults in power transformers by dissolved gas-in-oil chromatographic analysis: Doble Client Conference. 42 A.I.C. 75 Section 10.2.

15.44   CIGRE 1975, Detection of, and research for, the
          characterisation of incipient faults from analysis of
          dissolved gas in the oil of insulation. ELECTRA, 42,
          31.

15.45   IEC 599.   Interpretation of the analysis of gases in
          transformer and other oil-filled electrical equipment
          in service.

15.46   Belanger, G. and Missout, G., 1980, Electrochemical
          determination of hydrogen dissolved in transformer
          oil.   Analytical Chemistry 52, 2406.

15.47   Belanger, G. Missout, G. and Gibeault, J.P., 1980,
          Laboratory testing of a sensor for hydrogen dissolved
          in transformer oil.   IEEE Trans. EI 15, 144.

15.48   Graham, J., 1982, On-line analyser monitors faults
          in oil-filled units.   Electrical Review, 211,
          No. 18, 29.

15.49   Wilson, A.   To be published.

15.50   Hodge, J.M., Miller, T., Roberts, A. and Steel,J.G.,
          1982, Generator monitoring systems in the U.K.
          CIGRE Paper 11.08.

15.51   Wood, J.W., 1982, Condition monitoring of turbo-
          generators.   Electronics and Power, 28, 682.

15.52   Barton, S.C., Gibbs, E.E. and Kostoss, J.M., 1981,
          Generator gas monitoring system and detection of
          stator core hot spots by infra red camera. E.P.R.I.
          Conference and Workshop, WS-80-133.

Index

Individual materials are grouped under Gaseous insulants, Liquid insulants or Solid insulants, not under alphabetic name.